To —

— because he
no intention of
buying one —

David.
June 1972.

Elements of Estimating

Elements of Estimating

David S. M. Hall here provides students of Estimating, Quantity Surveying and Building Management with the fundamentals of the complex art of estimating for building work.

The book covers the examination requirements of the Institute of Building Final Part 1 and the Estimating examinations of the Royal Institution of Chartered Surveyors and the Institute of Quantity Surveyors.

The 21 chapters include: Introduction to Price Analysis; The Labour Rate; Excavation and Earthwork; Concrete Work; Brickwork and Blockwork; Rubble Walling; Roofing; Carpentry; Joinery; Structural Steelwork and Metalwork; Plumbing; Plasterwork and Finishings; Glazing; Painting and Decorating; Drainage; Preliminaries; Completion of the Tender; Approximate Estimating; The Examination Approach; Examination Papers; Basic Prices of Materials.

At the end of each chapter examples of typical price build-ups are given. These reflect the orderly approach and layout which the author advises and adopts.

David S. M. Hall, AIQS, AIArb, was for ten years a Quantity Surveyor with John Laing Construction Limited and has for twenty years been an Estimator with this well-known company. He also lecturers to post-graduate building management trainees at The London Polytechnic, School of Architecture.

Elements of Estimating

David S. M. Hall

B. T. Batsford Limited

'MONY A MICKLE MAKS A MUCKLE...'
Robert Burns

© David S. M. Hall 1972

First published 1972

ISBN 0 7134 0526 0 (hard cover)
ISBN 0 7134 0527 9 (paper back)

Filmset by Keyspools Limited, Golborne, Lancashire
Printed and bound in Great Britain by
C. Tinling & Co. Limited, Prescot, Lancashire
for the publishers
B. T. BATSFORD LIMITED
4 Fitzhardinge Street, London W1H 0AH

Contents

Preface		6
1	Introduction to Price Analysis	7
2	The Labour Rate	9
3	Excavation and Earthwork	13
4	Concrete Work	28
5	Brickwork and Blockwork	42
6	Rubble Walling	51
7	Roofing	52
8	Carpentry	57
9	Joinery	60
10	Structural Steelwork and Metalwork	64
11	Plumbing	65
12	Plasterwork and Finishings	68
13	Glazing	74
14	Painting and Decorating	76
15	Drainage	78
16	Preliminaries	81
17	Completion of the Tender	91
18	Approximate Estimating	93
19	The Examination Approach	95
20	Examination Papers	98
21	Basic Prices of Materials	104
Index		107

Preface

This volume has been written to provide students of Estimating, Quantity Surveying and Building Management with the simple fundamentals of the complex art of estimating for building work.

It is not intended to be a manual for professional estimators whose needs are adequately catered for in existing publications. These admirable volumes of reference, however, with their multitudinous columns of labour and material constants have proved to be a source of dismay to the student preparing for his professional examination.

The basic layout of the book was prepared from the author's notes on a course of lectures given to post-graduate building management trainees at the London Polytechnic School of Architecture. The examination was the Institute of Building Final Part 1, but the text equally covers the requirements of the Estimating examinations of the Royal Institution of Chartered Surveyors and the Institute of Quantity Surveyors.

The labour and plant outputs used are average outputs achievable under normal conditions. It is part of the practising Estimator's expertise to vary normal outputs according to the particular conditions of the actual project being priced. There is, however, a vast difference in approach between an estimator pricing a bill of quantities for tender purposes and a student tackling an examination paper. In the latter case, neither the relevant quantities nor details of the location of the work are provided and it is precisely these details which affect the judgment of the skilled estimator. For this reason, many of the sophistications adopted in practice are not available to the examinee, which, from one point of view, makes his task somewhat easier.

Students are advised not to attempt to memorise the labour outputs in parrot fashion, but rather to endeavour to obtain an overall appreciation of the range of outputs relevant to each trade.

At the end of each chapter examples of typical price build-ups are given. A study of these will reveal the orderly method of approach and layout to be adopted at all times. Students should also endeavour to familiarise themselves with the *National Working Rule Agreement* issued by the National Joint Council for the Building Industry, 11 Weymouth Street, London W1, and with the *Code of Estimating Practice*, published by the Institute of Building, 10B Anglemere, King's Ride, Ascot, Berkshire.

Both these publications contain essential information for students preparing for an examination in Estimating.

1 Introduction to Price Analysis

In building up prices the student should adopt a methodical technique which is not likely to let him down under the stress of examination conditions.

The basic approach to every price analysis should be concise and uniform, each build-up comprising the cost of the following components:

1 Materials
2 Unloading
3 Waste
4 Labour
5 Plant
6 Sundry items

1 *Materials*

In practice, the estimator obtains quotations for the various materials required from merchants or suppliers. Normally, the prices quoted include for delivery to the site and the estimator bases his build-up on the most favourable quotation for each item.

In an examination, the cost of the relevant materials is provided. Extreme care must be exercised in ensuring that the units are correctly read and interpreted. For example, sand and aggregates can be priced either in tonnes or cubic metres and the student must not confuse the two.

2 *Unloading*

An allowance for unloading on site must be made for most materials, including bagged cement, bricks, reinforcing rods and precast concrete goods. Sand and aggregates can normally be tipped in a position adjacent to the concrete mixer and require no unloading.

3 *Waste*

An allowance for waste is added to the cost of most materials used, generally a percentage of the supply price varying between $2\frac{1}{2}\%$ and 15%. Waste occurs after the materials have been unloaded and stored on site. The percentage allowance is therefore applied to the cost of the materials delivered to site plus the cost of unloading.

4 *Labour*

The labour outputs adopted in the text are average outputs achievable under normal conditions.

In practice, labour outputs vary considerably depending on the size of the contract, on site conditions, on the weather, on the difficulty or ease of working and on the quality of labour in the particular area.

Outputs vary in different parts of the country. Outputs vary from one man to another. The output of the same man varies from day to day.

The art of the skilled estimator is to assess all these variables to arrive at a competitive price.

5 Plant

Outputs achieved with mechanical plant vary in their extremes to a greater extent even than labour outputs and depend upon the actual equipment used, its utilisation factor and the quantity of work to be carried out. Usually if mechanical plant can be efficiently utilised it is cheaper than hand labour. The range of plant covered in the text includes mechanical excavating equipment, concrete mixers, dumpers, lorries, rollers, pumps, mechanical hoists and tower cranes.

6 Sundry items

Under this heading are considered the ancillary costs which must not be forgotten in a price analysis.

These are the minor items, often of negligible cost, which are nevertheless essential components of the finished job, for example, the binding wire and plastic spacers in reinforcement, the yarn in drain pipe joints and the nails in carcassing timber. Many of these essential items are ignored by examinees and it is often the correct application of the sundry items which indicates to an examiner that the student has a complete understanding of the construction involved.

2 The Labour Rate

In preparing a tender for a building contract one of the first tasks of the estimator is to calculate the current gross hourly rates for tradesmen and labourers to be used throughout the pricing of the Bills of Quantities.

The various components of the gross labour rates are as follows:

1 *Basic wage rate*

The basic wage rates for tradesmen and labourers are fixed by the National Joint Council for the Building Industry and are published in the *National Working Rule Agreement* and in the technical press.

2 *Inclement weather*

In the event of adverse weather conditions sufficient to stop work, operatives are paid a guaranteed weekly minimum wage subject to their reporting on site and being available during the normal working hours each day.

They receive half pay for each hour lost due to inclement weather and should the total, plus the hours worked be less than 40, the employer makes the difference up to 40 hours.

An addition of 2% to the basic labour rate is normally sufficient to cover the contractor's cost of guaranteed time payments.

3 *Non-productive overtime*

Additional payment is made to operatives for overtime working. Assuming a five day working week, Monday to Friday, extra overtime payments are as follows:

First hour	—Time-and-a-quarter
Next two hours	—Time-and-a-half
Thereafter	—Double time
Saturdays and Sundays	—Up to 4.00 p.m.: time-and-a-half
	Thereafter: double time

The extra time payments are made at the basic wage rate and if, for example, the working week is five days of nine hours (the basic week being five days of eight hours), the non-productive overtime is five days at $\frac{1}{4}$ hour = $1\frac{1}{4}$ hours per man per week.

4 *Sick pay*

Provision for the number of days lost due to sickness is generally covered by an insurance scheme. An average allowance is £0·10 per man per week.

5 Trade supervision and gangers

The normal proportion is taken as one trade foreman supervising eight to twelve tradesmen. Gangers also are included in the ratio of one to every eight to twelve labourers, their full time being regarded as supervisory.

This item is priced in the example shown on page 12, by the addition of $12\frac{1}{2}\%$ in the wage rate build-up.

6 Tool money

Craftsmen are paid a weekly allowance for the provision, maintenance and upkeep of their working tools, the amounts being laid down in the *National Working Rule Agreement*. An average of £0·20 per week is included for this item in the gross rate build-up, a similar amount being included for labourers to cover minor allowances such as boot money and dirt money.

7 Training

To provide for the cost of the Construction Industry Training Board Levy and training of operatives within the company, allow 2%.

8 National Insurance

National Insurance payments by contractors comprise a flat rate for every man employed and a graduated rate depending on his gross pay.

9 Selective Employment Tax

This tax was halved, to £1·20 per man week, in June 1971, and the Government has announced that it will be abolished in June 1973, when a form of Value Added Tax is expected to be introduced.

10 Holidays with pay

The current employer's contribution is £1·25 per man/week for annual holidays and £0·34 and £0·29 for public holidays for tradesmen and labourers respectively. This provides operatives with three weeks annual holiday and four days public holiday.

11 Severance pay and sundry costs

An allowance of 1% is made for the following items:
  Severance pay
  Loss of production during notice
  Absenteeism
  Turnover of labour

12 *Employer's liability and third party insurance*

Employer's liability is sometimes referred to as workmen's compensation. Employers protect themselves by insurance against claims for injury or accidents to operatives during the course of their employment.

Third party insurance is the employer's insurance against injury or accident to the public or to adjoining property.

Premiums are usually based upon the annual value of wages paid by the employer and at present rates an addition of $1\frac{1}{4}\%$ provides for the necessary insurance.

13 *Extra payments*

Under the *National Working Rule Agreement* operatives performing certain tasks involving discomfort, inconvenience or risk, such as working at great heights or working under abnormally wet or dirty conditions receive extra hourly payments. In addition operatives whose work involves continuous extra skill or responsibility, such as timbermen, scaffolders, and compressed air tool operators receive extra payment.

These allowances are detailed in the *National Working Rule Agreement* and are added in the text to the operative's rate where they apply.

Gross hourly rates

There is no universally adopted method for compiling the gross hourly rates for tradesmen and labourers. Individual estimators adopt the system which suits them best and whilst some include the cost of travelling time and fares, non-productive overtime and trade supervision in the gross labour rates, others prefer to price one or all of these items in the Preliminaries.

The *Code of Estimating Practice,* published by the Institute of Building, contains an example of gross rate build up calculated on an annual basis which is subsequently converted to an hourly rate.

The build up which follows is a typical example of the gross labour rate calculated on a weekly basis and converted to an hourly rate.

It is usual in Estimating examinations for the gross labour rates which are to be used in answering questions to be stated in the paper.

Candidates should, however, be familiar with the various items added to the basic rates of wages to produce the gross rates used in price build-ups.

THE LABOUR RATE

Gross hourly rates
Typical build-up of gross hourly rates

		Tradesmen £		Labourers £
(a)	Basic rate—45 hours @ £0·50½	= 22·73	£0·43	= 19·35
(b)	Inclement weather—2%	= 0·45		= 0·39
(c)	Non-productive overtime —1¼ hours @ £0·50½	= 0·63	£0·43	= 0·54
(d)	Sick pay	= 0·10		= 0·10
		= 23·91		= 20·38
(e)	Trade supervision—12½%	= 2·99		= 2·55
			NWR	
(f)	Tool money	= 0·20	*Allowances* =	0·20
		= 27·10		= 23·13
(g)	Training and CITB Levy—2%	= 0·54		= 0·46
(h)	National Insurance:			
	Flat rate	= 0·89		= 0·89
	Graduated	= 1·22		= 0·97
(i)	Selective Employment Tax	= 1·20		= 1·20
(j)	Holidays with pay:			
	Annual	= 1·25		= 1·25
	Public	= 0·34		= 0·29
		= 32·54		= 28·19
(k)	Severance pay—1%	= 0·33		= 0·28
		= 32·87		= 28·47
(l)	Employer's liability—1¼%	= 0·41		= 0·36
	Per week	= <u>33·28</u>		= <u>28·83</u>
	Divide by 45 = per hour	= <u>£0·74</u>		= <u>£0·64</u>

3 Excavation and Earthwork

The Unit of Measurement is the Metre

Measurement of excavation

Excavation is measured as the net volume excavated and is generally described as 'excavate and get out'. The subsequent disposal of the excavated material either follows in the same item or appears later in the Bills of Quantities as a separate item of removal.

The cost of excavation is affected by such factors as the quantities involved, the depth below ground level, whether or not a mechanical excavator can be used and the nature of the soil relative to the ease of digging it out.

Bulking of excavated material

Soil increases in bulk from 25% to $33\frac{1}{3}$% when excavated and deposited in spoil heaps or when loaded into lorries and account must be taken of this bulking factor in subsequent handling and disposal of the material off site.

Hand excavation

Labour outputs in firm soil

Excavate and deposit or load into barrows

	Hours per m^3
Surface excavation up to 300 mm deep	$2\frac{1}{2}$
Excavate to reduce levels	2
Excavate trenches not exceeding 1·50 m deep	3
Excavate basements ditto	$2\frac{1}{2}$
Excavate small pits, column bases ditto	4
For each additional 1·50 m depth ADD	$1\frac{1}{2}$
Extra for loading spoil into lorries ADD	$1\frac{1}{2}$

Multipliers for differing soils	*Cost multiplier*
1 Loose soil, loam or dry sand	0·67
2 Firm soil or ordinary clay	1·00
3 Stiff or heavy clay, gravel or loose chalk	1·50
4 Soft rock or solid chalk	3·00
5 Hard rock, stone or concrete	5·00

Disposal of excavated material

Excavated material is disposed of on the site if it can be used to make up levels, otherwise it must be removed off site to a tip. If the material can be used on site the relevant item in the Bill of Quantities will read, for example, 'remove a distance not exceeding 200 metres and deposit in spoil heaps'.

Hand excavated material can often be removed to its final location in barrows. This involves loading and wheeling and the maximum economic length of wheel is taken to be about 100 metres. Exceeding 100 metres it is often cheaper to load the spoil into dumpers or lorries and transport it to the required location. If the spoil is not required on site, the relevant item will read 'remove excavated material off site to tip'. In this instance it becomes the contractor's responsibility to find a tip, but in practice, firms supplying aggregates will frequently quote a price for removing the surplus spoil on the return journey of their lorries to the pits.

Labour outputs	*Hours per m³*
Wheel excavated material 25 metres and deposit	½
For each additional 25 metres ADD	½
Return fill and ram around foundations	1
Spread and level in layers 150 mm thick	½

Mechanical excavation

Types of excavating equipment

Mechanical excavators	*Use*
(a) Skimmer	General excavation down to the level of its own tracks, generally not exceeding 1·50 m deep.
(b) Face shovel	Excavation at faces over 1·50 m deep.
(c) Dragline	Bulk excavation below track level such as basements and reducing levels.
(d) Back acter	Drain and surface trench excavation.
Crawler tractor shovels	General site excavation and loading to lorries for subsequent disposal. Bucket sizes from ½ to 1½ cubic metres.
Tractor-mounted shovels	Oversite and surface trench excavation.
Bulldozers	Reducing levels oversite, filling and grading, forming embankments.
Tractor scrapers	Bulk surface excavation and removal. Scraper capacity from 4 to 12 cubic metres.

EXCAVATION AND EARTHWORK

Utilisation of excavating equipment

Mechanical excavating equipment cannot normally be usefully employed all the time it is on site. Allowance must be made for non-productive time standing and for travelling.

Depending upon the quantities of excavation, utilisation of mechanical equipment can vary from as little as 30% to as much as 95%. A reasonable average is 75% which implies that for three-quarters of its time on site the equipment is usefully working and for one quarter of its time is non-productive.

Whilst the machine is non-productive its hire rate and the cost of the driver have to be paid by the contractor and a percentage addition to the net machine rate is made to cover this non-productive time.

If a machine is usefully operated only 50% of the time, its net rate must be doubled to arrive at the gross operating rate. Similarly, if it is operating 75% of the time, $33\frac{1}{3}$% must be added to the net rate to arrive at the gross rate.

The main machine utilisation percentages are as follows:

Estimated utilisation	Addition to net rate
50%	100%
66%	50%
75%	33%
80%	25%
90%	11%

Hire rate for mechanical plant

To calculate the hire rate of mechanical plant the following costs must be taken into consideration:
1. Capital cost
2. Depreciation
3. Interest on capital outlay
4. Maintenance and repairs
5. Insurance and licence
6. Residual value
7. Transport to and from site

The last item is normally priced in the Preliminaries as a lump sum covering all the plant required on site.

An estimate is made of the probable life of the machine and the number of working hours per annum to be expected from it. A reasonable working life for site plant can be taken as from three to five years depending on the machine and the number of working hours per annum as follows:

Lorries and dumpers	2200 hours
Other mechanical plant	1800 hours

EXAMPLE

Assume capital cost £3000, 3 year use, working 1800 hours per annum and residual value £900.

		£
Capital cost	=	3000
Interest at 5% per annum	=	450
Maintenance and repairs 10% per annum on capital cost	=	900
Insurance and licence say £50 per annum	=	150
		4500
Less Residual value	=	900
	=	3600
÷ 3 years	=	£1200
÷ 1800 hours. Hire rate per hour	=	£0·67

In an estimating examination the hire rate of the plant to be used is generally given, including the cost of fuel and lubricating oil. The student converts this rate into a gross operational hourly rate by adding the cost of the driver, a banksman if required, and the utilisation factor.

Gross hourly rate for a mechanical excavator (0.53 m^3 bucket)

		Per hour £
Hire rate of machine including fuel	£	= 1.30
Driver:		
Gross labourer's rate	= 0·64	
NWR plus rate	= 0·06	
Maintenance and starting time 25% on basic rate £0·43 and plus rate = £0·49	= 0·12	
	= 0·82	= 0·82
Banksman		
Gross labourer's rate	= 0·64	
NWR plus rate	= 0·02	
	= 0·66	= 0·66
		= 2·78
Machine utilisation 75% therefore add $33\frac{1}{3}$%		= 0·93
Gross machine rate		= 3·71

NOTES

1. The NWR plus rates above refer to the *National Working Rule Agreement*.
2. A banksman should be provided for all mechanical excavating equipment.

Outputs of mechanical excavators

Basic outputs

Excavator type	Bucket size m^3	m^3 per hour
10 RB	0·28	7
17 RB	0·48	9
19 RB	0·53	12
22 RB	0·62	15

The basic outputs apply to:
1. Excavating oversite up to 500 mm deep
2. Excavating basements

They should be increased by 50% for bulk excavation to reduce levels, i.e. to 10, 14, 18 and 22 m^3 per hour.

They should be reduced by $33\frac{1}{3}$% for excavating drain and surface trenches, i.e. to 5, 6, 8 and 10 m^3 per hour respectively.

Crawler tractor shovels

A crawler shovel will excavate oversite and load to lorries 15 to 20 cubic metres an hour.

Scrapers

Outputs of tractor-powered scrapers vary tremendously depending on the total volume of excavation to be carried out, the size of scraper and the distance of haul. A seven cubic metre machine will excavate to reduce levels, remove up to 400 metres and spread and level between 30 and 40 cubic metres an hour.

Disposal of machine excavated material

NOTES

1. Loading the excavated material into lorries is covered by the outputs taken.
2. Normal soil bulks about $33\frac{1}{3}\%$ when excavated. Three cubic metres of solid material therefore bulks to become 4 cubic metres loose when loaded into lorries.
3. A 5 tonne tipper lorry carries 3 cubic metres of solid material.

The estimator has to assess the number of lorries or dumpers necessary to remove the excavated material at such a rate as to keep the excavating machine economically working.

Assuming that the machine is excavating at an average output of 12 cubic metres an hour, each lorry will be fully loaded with 3 cubic metres of solid material (4 cubic metres loose) in fifteen minutes.

An economic work cycle follows if the lorry travels to the spoil heap, discharges its load and returns to the excavator in a further fifteen minutes. Such a cycle requires two lorries. If the return journey to and from the tip takes thirty minutes, three lorries are required to keep the machine adequately employed.

EXCAVATION AND EARTHWORK 19

The proportion of hand excavation and mechanical excavation

The estimator has to assess on each item of excavation being priced, the expected proportion of hand excavation to machine excavation. It is unlikely that a machine will be able to excavate 100% of any particular item. Labour trimming of the sides and bottom is generally required and hand excavation may be necessary in areas of difficult access for plant. This is a matter of estimating judgment.

In pricing basement excavation the estimator may visualise 90% machine digging and 10% hand digging. In surface trenches the ratio could be 50%–50%.

The calculation for a 75% machine to 25% hand excavated item is as follows:

Mechanical excavation

Excavator £3·71 per hour
÷ 12 m^3 per hour   = £0·31 m^3

Hand excavation

3 hours labourer @ £0·64   = £1·92 m^3

75% by machine @ £0·31   = £0·23
25% by hand @ £1·92   = £0·48

                    per m^3 = £0·71

NOTE

A certain proportion of hand excavation is generally necessary in all predominantly machine excavated items.

Analyses

Hand Excavation **Firm soil**

1 *Excavate to remove topsoil average 150 mm deep and wheel 50 metres to spoil heap*   Per square metre
   £

   Excavate $2\frac{1}{2}$
   Wheel   1
              ___
              $3\frac{1}{2}$ hours

                                              £
   $3\frac{1}{2}$ hours labourer @ £0·64   = 2·24 m³

   $150 \text{ mm} = \dfrac{150}{1000} =$ per m²   = 0·34

2 *Excavate to reduce levels and get out*   Per cubic metre
                                              £
   2 hours labourer @ £0·64 =   per m³   = 1·28

3 *Ditto in heavy blue clay*   Per cubic metre
                                              £
   2 hours labourer @ £0·64   = 1·28
   Multiplier for heavy clay   = 1·50   = 0·64
                                  per m³   = 1·92

4 *Excavate surface trench exceeding 1·50 m not exceeding 3 m deep and get out*   Per cubic metre
                                              £
   $4\frac{1}{2}$ hours labourer @ £0·64 =   per m³   = 2·88

5 *Wheel excavated material 50 m and spread and level in layers 150 mm thick*   Per square metre
                                              £
   Wheel 50 m   1
   Spread and level $\frac{1}{2}$
                      ___
                      $1\frac{1}{2}$ hours

                                              £
   $1\frac{1}{2}$ hours labourer @ £0·64   = 0·96 m³
   $150 \text{ mm} = \dfrac{150}{1000} =$ per m²   = 0·14

6 *Level and ram bottom of excavation to receive hardcore*   Per square metre
                                              £
   $\frac{1}{10}$ hour labourer @ £0·64 =   per m²   = 0·06

EXCAVATION AND EARTHWORK

Mechanical excavation

7   *Excavate to remove topsoil average 200 mm*   Per square metre
    deep and get out   £
    Assume gross cost of mechanical excavator
    £3·71 per hour
    Mechanical excavator £3·71 ÷ 12 m³ per hour
    $\qquad\qquad\qquad\qquad\qquad = £0\cdot31\ m^3$
    200 mm thick $= \dfrac{200}{1000} =$ per m² $= \underline{\underline{0\cdot06}}$

8   *Excavate to reduce levels and get out*   Per cubic metre
    Mechanical excavator £3·71 ÷ 18 m³ per hour   £
    $\qquad\qquad\qquad\qquad = $ per m³ $= \underline{\underline{0\cdot21}}$

9   *Ditto load to lorries, remove 400 metres and*   Per cubic metre
    deposit in spoil heap   £
    Excavate as item 8 $= 0\cdot21$
    Lorries—2 @ £1·55 per hour
       $= £3\cdot10 \div 18$ m³ per hour $= 0\cdot17$
    Attend spoil heap $= 0\cdot03$
                                       per m³ $= \underline{\underline{0\cdot41}}$

In the foregoing items it has been assumed that all the excavation is carried out by machine. Note that the spoil is loaded to lorries as part of the output.

The two analyses which follow introduce a proportion of hand excavation.

10 *Excavate to reduce levels, load to lorries,*   Per cubic metre
    remove 400 metres and deposit in spoil heap   £
    Assume 75 % mechanical excavation
       and 25 % hand excavation
    Mechanical excavation and lorries as Item 9
       £0·38 per m³ × 75 % $= 0\cdot29$
    Hand excavation   2
    Load to lorries $1\frac{1}{2}$
                            $\overline{3\tfrac{1}{2}}$ hours
    $3\tfrac{1}{2}$ hours labourer @ £0·64 = £2·24 m³ × 25 % $= 0\cdot56$
    Attend spoil heap $= 0\cdot03$
                                       per m³ $= \underline{\underline{0\cdot88}}$

11 *Excavate surface trench not exceeding*   *Per cubic metre*
    1·50 metres deep and get out   £
    Assume 50% mechanical excavation
      and 50% hand excavation
    Mechanical excavator £3·71 ÷ 8 m³ per hour
        = £0·46 per m³ × 50%   = 0·23
    Hand excavation 3 hours labourer @ £0·64
        = £1·92 m³ × 50%   = 0·96
                                per m³   = <u>1·19</u>

NOTES

a   Observe the variation in the rate according to the estimator's judgment of the proportion of machine to hand excavation.
b   The procedure is always the same:
     (i) Calculate the cost of machine excavation.
     (ii) Calculate the cost of hand excavation.
     (iii) Apportion the percentage of each.

12 *Excavate oversite to reduce levels, remove*
    800 metres and deposit, spread and level to   *Per cubic metre*
    make up levels   £
    Assume a scraper can be used,
        gross cost £5·50 per hour
    Scraper £5·50 ÷ 35 m³ per hour   per m³   = <u>0·16</u>

13 *Ditto using a crawler tractor shovel and*   *Per cubic metre*
    lorries   £
    Crawler shovel £2·75 ÷ 18 m³ per hour   = 0·15
    Assume lorry cycle as follows:

                                     Minutes
       Load 3 m³ @ 18 m³ per hour   10
       Travel   6
       Deposit   5
       Return   6
       Margin   <u>3</u>
                                                 30

Since the loading is ⅓ of the total cycle, 3 lorries are required.
3 lorries @ £1·55 = £4·65 per hour
    ÷ 18 m³ per hour   = 0·26
Spread and level ⅙ hour labourer @ £0·64   = 0·11
                                         per m³   = <u>0·52</u>

Disposal of excavated material

14 *Remove excavated material 400 metres to spoil heap*
    Per cubic metre
    £

Loading time at average 12 m^3 per hour
= 15 minutes.

Travel	400 m	4
Deposit		5
Return	400 m	4
Margin		2
		15 minutes

An economic cycle is achieved with two lorries, one being loaded, the other travelling to spoil heap.

2 lorries @ £1·55 per hour = £3·10 ÷ 12 m^3
    (average output)   = 0·26
Attend spoil heap   = 0·03
                                      per m^3 = <u>0·29</u>

NOTE

The allowance for attending the spoil heap is for the cost of providing a man to direct the lorries or for occasional bulldozer attendance to level off the spoil.

15 *Extra for removing each additional 400 metres*
    Per cubic metre
    £

The extra cost is in the additional lorry time to travel the extra distance. Say the extra time is 6 minutes.

Then extra cost = $\frac{1}{10}$ hour @ £1·55
                   = <u>£0·16</u> per load.
The lorry load is 3 m^3 solid
                 = £0·16 ÷ 3 = per m^3 = <u>0·05</u>

16 *Remove surplus excavated material to tip off site*
    Per cubic metre
    £

An aggregate supplier or a haulage contractor will often quote for removal of spoil off site. Assume a quotation of £0·54 per cubic metre has been received.

Every 3 m^3 of compacted soil will become 4 m^3 of loose material for which the haulier will be paid £0·54 m^3 = £2·16 per load.

The contractor, however, will be paid for removing only 3 m^3, measured in the solid. He must accordingly increase the haulier's rate by

$33\frac{1}{3}$ to a bill rate of £0·72 m³. He will then be paid 3 m³ @ £0·72 = £2·16 per load.

If no quotation for removal of excavated material is obtained, the contractor has to find his own tip and the build-up is as follows:

Assume a tip has been located at 8 km distance:

	Minutes	
Load lorries	15	
Travel 16 km	30	
Deposit	5	
Margin	10	
	60	£
1 hour lorry @ £1·55 ÷ 3 m³		= 0·52
Attend tip		= 0·03
Tipping fee		= 0·10
	per m³	= 0·65

NOTE

The tipping fee, if any, must not be forgotten. This money is paid to the owner of the tip.

Planking and strutting

Planking and strutting is measured in square metres of actual face upheld. Having priced the item in the Bills, the contractor is paid the rate inserted whether or not he may subsequently require to install heavier boarding than he assumed or whether or not he uses planking and strutting at all. The responsibility for upholding the face of the excavation is his. So is the risk of failing to do so.

The cost of planking and strutting is dependent upon the type of subsoil, the amount of water in the ground and the length of time the excavation is likely to remain open.

In dry conditions and firm ground very little planking and strutting may be required. In wet, loose soil, close boarding may be necessary.

Planking and strutting to basements and similar deep excavations is necessarily heavier in the face material and the strutting than planking and strutting to drain trenches or foundation trenches and can cost from two to four times as much per square metre.

A timberman is paid a NWR allowance of £0·01 per hour above a labourer's rate.

MEMORANDA

Planking and strutting to surface trenches

Depth	Close boarding Timber m^3	Per square metre Timberman hours
Not exceeding 1·50 m	$\frac{1}{24}$	$\frac{1}{3}$
Not exceeding 3·00 m	$\frac{1}{18}$	$\frac{1}{2}$
Not exceeding 4·50 m	$\frac{1}{12}$	$\frac{3}{4}$

NOTE

Timber used for planking and strutting is normally used to destruction and an average life expectancy is 12 uses.
The material cost must be divided by 12 to arrive at a cost per use.
For open boarding, sometimes called 'hit and miss' timbering, use half the foregoing timber and labour.

Analyses

Planking and strutting

1 *Planking and strutting to sides of surface*   *Per square metre*
  trenches not exceeding 1·50 m deep   £
  Assume close boarding is required.   £

Carcassing timber per m^3	= 28·00
Unload 2 hours @ £0·64	= 1·28
Waste 10%	= 2·93
Spikes, dogs, wedges	= 0·40
per m^3	= 32·61

    Timber $\frac{1}{24}$ m^3 @ £32·61

per m^3	= £1·36	
Divide by 12 uses		= 0·11
Timberman $\frac{1}{3}$ hour @ £0·65		= 0·22
	per m^2	= 0·33

NOTE

The rate for open boarding, or 'hit and miss' boarding would be 50% of the above.

2 *Ditto not exceeding 4·50 m deep*   *Per square metre*
  Timber $\frac{1}{12}$ m^3 @ £32·51   £

per m^3	= £2·72	
Divide by 12 uses		= 0·23
Timberman $\frac{3}{4}$ hour @ £0·65		= 0·49
	per m^2	= 0·72

EXCAVATION AND EARTHWORK

Hardcore and ashes

MEMORANDA

Allow 25% consolidation on hardcore and 40% on ashes.

Analyses

1. *Hardcore in bulk filling*   *Per cubic metre*
   £
   Hardcore £1·30 per m³   = 1·30
   Consolidation 25%   = 0·33
   Labourer 1 hour @ £0·64   = 0·64
                            per m³ = $\underline{\underline{2·27}}$

2. *Hardcore in bed 150 mm thick spread and levelled under floors and blended with fine material including levelling and ramming ground under*   *Per square metre*
   £
   Hardcore as above £2·27 m³
   $\times 150$ mm thick $= \dfrac{150}{1000}$ = 0·34
   Blinding—say 25 mm thick allow 20%   = 0·07
   Level and ram ground:
      Labourer $\frac{1}{10}$ hour @ £0·64   = 0·06
      Rolling $\frac{1}{20}$ hour @ £0·64   = 0·03
                            per m² = $\underline{\underline{0·50}}$

NOTES

a. The blinding material is lost in the interstices of the stone and the full depth of hardcore required should be taken.
b. On large areas a mechanical roller would be used.

3. *Ashes in bed 100 mm thick spread and levelled under floors*   *Per square metre*
                                £   £
   Ashes £1·50 per m³   = 1·50
   Consolidation 40%   = 0·60
   Labourer $\frac{1}{2}$ hour @ £0·64   = 0·32
                 per m³ = $\underline{\underline{2·42}}$
   $\times 100$ mm thick $= \dfrac{100}{1000}$ = 0·24
   Level and ram ground as above   = 0·06
   Rolling ditto   = 0·03
                            per m² = $\underline{\underline{0·33}}$

4 Concrete Work

The Unit of Measurement is the Metre

Concrete mixes

Concrete mixes can be specified in two ways, by volume or by weight. When specified by volume the concrete is described as a 1:2:4 mix or a 1:3:6 mix, the first unit being the proportion by volume of cement in the mix, the second being the volume of sand and the third the volume of aggregate.

Concrete mixes can also be described as 1:8 or 1:10. This means that one part of cement by volume is to be mixed with 8, or 10, parts of 'all-in' ballast, a material containing the sand and aggregates already mixed together in the natural state.

Materials

Cement

Cement is supplied either in 50 kilogramme bags or in bulk. It weighs 1440 kg per cubic metre but to cater for waste allow 1500 kg or 1·5 tonnes per cubic metre.

Sand and aggregates

The price of sand and aggregates delivered to site can be quoted in tonnes or in cubic metres. To convert the price per tonne into the price per cubic metre, add 50%:

	£
Sand per tonne	= 1·60
add 50%	= 0·80
Sand per m³	= 2·40

Water

The cost of the water used in mixing concrete is included in the Preliminaries Bill, where the total cost of supplying water for the whole contract is calculated.

Unloading

The allowance for unloading bagged cement from the delivery truck and transporting to the cement store is 1 labourer hour per tonne. Sand and aggregates are tipped straight into bins near the mixer and require no manual unloading.

CONCRETE WORK 29

Shrinkage of materials

When water is added to a mixture of dry cement, sand and aggregate and the components mixed together the resultant volume of concrete is less than the original volume of dry materials. For example, 6 cubic metres of dry materials would produce 4 cubic metres of concrete. The shrinkage factor in concrete is 50% and this is added to the cost of dry materials to arrive at the cost of a cubic metre of concrete.

The application of the shrinkage factor must not be forgotten by examinees.

Waste

An estimator may allow a small percentage for waste of concrete should the particular circumstances of the project being priced warrant it. For examination purposes it is assumed that all concrete mixed will be usefully placed and no specific allowance for waste is made.

The cost of materials in 1 cubic metre of concrete 1:3:6 mix

		£
Portland cement delivered site per tonne	=	7·90
Unload 1 hour @ £0·64	=	0·64
per tonne	=	8·54
× 1·5 tonnes per m³	=	4·27
per m³	=	12·81
		£
1 m³ Portland cement	=	12·81
3 m³ sand @ £2·40	=	7·20
6 m³ aggregate @ £2·24	=	13·44
	=	33·45
Add 50% for shrinkage	=	16·73
	=	50·18
÷ 10 parts	=	5·02
Cost of materials = £5·02 per m³		

Machine mixing

Machine mixing is more economical than hand mixing and gives more consistent results.

From the various sizes and types of concrete mixers on the market, the estimator must select the correct one on which to base his pricing. It must have sufficient capacity to accommodate the peak outputs required by the project programme but it must not be so big as to be idle for most of its time.

In an Estimating examination the size of the mixer to be used is stated, otherwise hand mixing can be assumed. Average outputs of concrete mixers are as follows:

Mixer type	Output m^3 per hour
7/5	2
10/7	3
14/10	4
18/12	5

The hourly cost of the mixer including fuel is divided by the average hourly output to arrive at the cost per cubic metre of the concrete mixer:

14/10 mixer £0·40 per hour
÷ 4 m³ per hour output
Cost of concrete mixer = £0·10 per m³.

Hand mixing

Allow 5 hours of a labourer per cubic metre.

Transporting the concrete

After mixing, the concrete is transported to the required placing position. On small contracts, and if the distance is short, the concrete is wheeled in barrows. On larger contracts, or if the distance from the mixer to the required position warrants it, the concrete is transported in dumpers or in lorries with concrete skips, by bogey rail, by pumping or by tower crane. For the present purpose we assume that the concrete is transported in barrows.

Labour for mixing and transporting

A typical labour gang for a 10/7 mixer is:
                1 Ganger
                1 Mixer driver
                2 Loaders
                2 Barrowmen

These six men mix and transport the 3 cubic metres produced by the 10/7 mixer in an hour. A 7/5 mixer requires five men and a 14/10 seven men. The method of pricing the labour gang is shown in the analyses.

CONCRETE WORK

Placing

The time taken to place a cubic metre of concrete varies according to its location and the table shows the number of man-hours required to place and consolidate a cubic metre of concrete in a variety of positions.

Placing concrete

	Labourer hours per m^3
Unreinforced concrete	
Mass concrete in foundations and stanchion bases	1
75 mm ground slabs	3
150 mm ditto	2
Reinforced concrete	
Foundations and stanchion bases	$1\frac{1}{2}$
75 mm ground slabs	$3\frac{1}{2}$
150 mm ditto	$2\frac{1}{2}$
100 mm suspended floors or roofs	5
150 mm ditto	$4\frac{1}{2}$
200 mm ditto	4
Columns and suspended beams exceeding $0.10\ m^2$ sectional area	8
Ditto 0.05–$0.10\ m^2$ ditto	10
Ditto not exceeding $0.05\ m^2$ ditto	12
100 mm walls	7
150 mm ditto	5
200 mm ditto	$4\frac{1}{2}$
Staircases and landings	8

Vibrating

In pricing reinforced concrete, allowance must be made for the cost of a poker vibrator to obtain full consolidation of the concrete around the reinforcing rods. There is no additional labour cost since the placers operate the vibrators as part of their duties. The cost of providing a vibrator is about £$0.37\frac{1}{2}$ per hour. Allowing for two being used = £0·75 per hour ÷ 3 m^3 per hour = £0·25 per cubic metre.

Surface finishes	*Per square metre labour hours*
Tamper finish	Nil extra
Spade finish	$\frac{1}{12}$
Trowelled finish	$\frac{1}{4}$

Curing

An allowance for curing the surface must be included on all horizontal concrete surfaces such as ground beds, roads and suspended floors and on vertical wall surfaces.

The type of curing required is generally specified, for example, sand or hessian fabric to be spread over the concrete and kept continuously damp for seven days. Polythene sheeting is also used to retain the concrete moisture during setting.

An allowance of £0·05 per square metre should be added to provide for use and waste of materials in curing concrete and for labour placing and removing.

Ready-mixed concrete

When the volume of concrete required does not warrant the expense of the installation of a site mixing plant, ready-mixed concrete may be used. It can also be used to cater for the peak outputs beyond the capacity of the site mixer.

If the site is a congested one with restricted working space consideration should always be given to the economics of purchasing ready-mixed concrete in lieu of site mixing.

The cost per cubic metre delivered to site is dependent on the quality of the mix and the delivery distance but under present conditions ready-mixed concrete is often economical for sites within a reasonable distance of the supplier's batching plant.

The cost of transporting in wheelbarrows should be added if the ready-mix delivery lorries cannot discharge directly to the required placing position. The normal placing costs as for site-mixed concrete are then added.

It is normal practice to include a small percentage for waste on ready-mixed concrete, say 1 to 2%.

Hoists

In pricing multi-storey structures, the cost of tower cranes and hoists for lifting the concrete is generally included in the Preliminaries.

The crane is employed in lifting many different materials and it serves little purpose to endeavour to allocate crane time to each item. As an academic exercise it is possible, but at best the allocation can only be arbitrary.

The total cost of hoisting equipment is therefore calculated and included in the Preliminaries as a lump sum.

Analyses

1 *Mass concrete 1:2:4 in foundations*   *Per cubic metre*
   Materials   £
   Portland cement per tonne   = 7·90
   Unload   1 hour @ £0·64   = 0·64
                                    per tonne   = 8·54
            × 1·5 tonnes per m^3 = 12·81
   2 m^3 sand @ £2·40 m^3 =   4·80
   4 m^3 aggregate @ £2·24 m^3 =   8·96
                                                         26·57
   Add 50% for shrinkage   = 13·29
                                                         39·96
                                                              £
                       per m^3 = ÷ 7 parts   =   5·69
   Mixing and transporting
   10/7 concrete mixer
       £0·33 per hour ÷ 3 m^3 per hour   = 0·11
   Labour   1 ganger
            1 mixer driver
            2 loaders
            2 barrowmen
                                     £
            6 men @ £0·64   = 3·84
               NWR plus rates = 0·12
                                  3·96
                   ÷ 3 m^3 per hour   = 1·32
   Placing
       1 hour labourer @ £0·64   = 0·64
                                     per m^3 = 7·76

NOTE

Observe the three stages of the build-up:
        Cost of   Materials
                  Mixing and transporting
                  Placing
This sequence and layout should be adhered to in all concrete pricing.

2 *Reinforced concrete 1:2:4 in foundations*   *Per cubic metre*
                                                              £
   Build up as item 1   = 7·76
   Add   Cost of vibrator   = 0·25
         Extra labour reinforced concrete
         ½ hour at £0·64   = 0·32
                                     per m^3 = 8·33

3 *Reinforced concrete 1:2:4 in ground*   *Per square metre*
   slab 150 mm thick   £
   Materials as item 1   = 5·69
   Mixing and transporting as item 1   = 1·43
   Vibrator as item 2   = 0·25
   Extra labour reinforced concrete as item 2   = 0·32
   Placing   2 hours labourer @ £0·64   = 1·28
                                              per m³   = 8·97
              150 mm thick = $\frac{150}{1000}$ = 1·35
   Curing   = 0·05
                                              per m²   = <u>1·40</u>

NOTE

Rates for concrete measured in square metres are best built up on a cubic metre basis then converted to square metres of the appropriate thickness.

4 *Reinforced concrete 1:1½:3 in suspended*   *Per square metre*
   floor 200 mm thick   £
   Materials
   1 m³ Portland cement @ £12·81 m³ (as item 1)   = 12·81
   1½ m³ sand @ £1·60 per tonne = £2·40 m³   = 3·60
   3 m³ aggregate @ £1·49 per tonne = £2·24 m³   = <u>6·72</u>
                                                       = 23·13
           Add 50% for shrinkage   = <u>11·57</u>
                                                       = <u>34·70</u>
           ÷ 5½ parts   = 6·31
   Mixing and transporting
     14/10 mixer including fuel £0·40 per hour
           ÷ 4 m³ per hour output   = 0·10
   Labour 1 ganger
           1 mixer driver
           2 loaders
           <u>3</u> barrowmen   £
           7 men @ £0·64   4·48
              NWR extras   <u>0·13</u>
                              4·61
              ÷ 4 m³ per hour   = 1·15
   Placing 4 hours labourer @ £0·64   = 2·56
   Vibrator £0·75 ÷ 4 m³ per hour   = <u>0·19</u>
                                              per m³   = <u>10·31</u>

CONCRETE WORK 35

$\times$ 200 mm thick $= \dfrac{200}{1000}$ = 2·06

Curing   as item 3   = 0·05

per m^2 = 2·11

NOTE

Observe the conversion of sand and aggregates from tonnes to cubic metres. Failure to do this is a common mistake made by students. It is essential to operate in the correct *units.* Note also the effect of the larger size mixer on the price for mixing.

5 *Reinforced concrete 1 : 1½ : 3 in suspended beams*
   exceeding 0·05 m^2, not exceeding 0·10 m^2   *Per cubic metre*
   sectional area   £
   Materials   as item 4   = 6·31
   Mixing and transporting   ditto   = 1·25
   Vibrator   ditto   = 0·19
   Placing   10 hours labourer @ £0·64   = 6·40

   per m^3 = 14·15

6 *Labour spade finish to concrete beds*   *Per square metre*
                                                           £
   Labourer $\tfrac{1}{12}$ hours @ £0·64 = per m^2 = 0·05

Mild steel reinforcement

The Unit of Measurement is the Kilogramme

MEMORANDA

1. Reinforcing rods are quoted at a price per tonne delivered to site. Whilst the unit of measurement is the kilogramme, it is more convenient to build up the price on a tonne basis and convert to kilogrammes at the end.
2. Unloading 2 hours per tonne.
3. Waste and rolling margin $7\frac{1}{2}\%$.
4. A steelfixer is paid the tradesman's rate, less £0·01 per hour.
5. The fixing outputs apply to mild steel, high tensile or square twisted rods.
   They apply to rods fixed in any position, i.e. in floors, columns, suspended beams, etc.
6. Add $33\frac{1}{3}\%$ extra labour for fixing links, stirrups and rods in short lengths.

Rod reinforcement, per tonne
Diameter

	6 mm	10 mm	12 mm	16 to 20 mm	25 mm and over
Steelfixer hours					
Cut, bend and fix	90	80	70	60	50
Binding wire, kg	20	15	15	10	10

Mesh or fabric reinforcement   *Per square metre*

MEMORANDA

1. Mesh reinforcement is supplied in sheets or rolls and is measured in square metres.

	Hours
2. Unload mesh 2 kg per m²	$\frac{1}{80}$
6 kg per m²	$\frac{1}{40}$

3. Waste and laps—15%
4. Labour fixing mesh in ground beds and roads:

mesh 2 kg per m²	$\frac{1}{20}$
6 kg per m²	$\frac{1}{10}$

Fixing in foundations and suspended floors—add 50%
Fixing in walls   —add 100%

5. Raking cutting
   waste $\frac{1}{6}$ m² per linear metre
   labour $\frac{1}{6}$ hour ditto
6. Curved cutting—add 50% to last.

CONCRETE WORK 37

Analyses

Mild steel reinforcement

1   *12 mm diameter mild steel rod reinforcement cut,*   *Per tonne*
    bent and fixed in suspended floors   £

	£
12 mm reinforcing rods £60·00 per tonne	= 60·00
Unload 2 hours per tonne @ £0·64	= 1·28
	= 61·28
Waste $7\frac{1}{2}\%$	= 4·60
Steelfixer 70 hours @ £0·73	= 51·10
Binding wire 15 kg @ £0·06 kg	= 0·90
Plastic spacers 200 @ £0·01 each	= 2·00
per tonne	= 119·88

NOTES

a   Observe that the waste is added to the cost *after* unloading.
b   If the price is required in kg, divide by 1000.

2   *6 mm diameter ditto*   *Per tonne*
                                                                      £

	£
6 mm reinforcing rods £65·50 per tonne	= 65·60
Unload 2 hours per tonne @ £0·64	= 1·28
	= 66·78
Waste $7\frac{1}{2}\%$	= 5·01
Steelfixer 90 hours @ £0·73	= 65·70
Binding wire 20 kg @ £0·06 kg	= 1·20
Plastic spacers 200 @ £0·01 each	= 2·00
per tonne	= 140·69

3   *6 mm ditto in links and stirrups*   *Per tonne*
                                                                                  £

	£
As item 2 per tonne	= 140·69
Extra labour links and stirrups $33\frac{1}{3}\%$ on £65·70	= 21·90
per tonne	= 162·59

Note the extra labour in fixing links and stirrups.

4   *Mesh fabric reinforcement 6·17 kg per m^2*   *Per square metre*
    in concrete roads   £

	£
Mesh fabric 6·17 kg £0·40 per m^2	0·40
Unload $\frac{1}{40}$ hour @ £0·64	0·02
	0·42
Waste and laps 15%	0·06
Steelfixer $\frac{1}{10}$ hour @ £0·73	0·07
Binding wire	0·01
per m^2	0·56

Formwork

The Unit of Measurement is the Metre

The pricing of formwork appears to present more difficulty to students than almost any other trade. Yet the fundamentals are quite straightforward and the build-ups should hold no terrors once the basic approach is understood.

The sequence in arriving at a price for formwork is as follows:

1. A certain volume of timber, sufficient to produce one square metre of formwork is priced.
2. Carpenters' time is allowed for fabricating the formwork panels from this timber.
3. The resultant total cost is divided by the assessed number of uses.
4. A percentage for waste and maintenance is added.
5. The formwork is fixed and struck.
6. Allowance is made for the sundry items—bolts, oil, wedges, struts, column cramps, etc.

Taking each item in turn:

1. The volume of timber required to produce one square metre of formwork depends primarily on the design of the formwork. The timber is quoted as a price per cubic metre, to which must be added the cost of unloading on site, waste in conversion and nails.
2. For fabricating the bulk timber into the required shape of formwork panel allow 36 hours of a carpenter per cubic metre of timber.
3. The Estimator assesses the number of uses likely to be obtained by studying the drawings and the programme of the project. Since these are not available to the student examinee the number of uses is normally given, otherwise a figure must be assumed.
4. Allow 10% for waste and maintenance following each use.
5. A typical formwork fixing gang comprises 1 carpenter and 2 labourers and the average hourly rate is calculated thus:

	£
1 carpenter @ £0·74	= 0·74
2 labourers @ £0·64	= 1·28
	= 2·02
÷ 3 =	0·67 per hour

This represents the average labour rate to be adopted for fixing and striking formwork.

6 It is important that the sundry items be included in the build-up. These show the examiner that the student is aware of the components of the finished formwork.

Wrought formwork

Formwork is described in the Bill of Quantities as being either 'sawn formwork' or 'wrought formwork'. When wrought formwork is specified, a smooth or fair faced finish is required to the concrete and an extra cost of £0·30 per square metre should be added. This provides for smooth timber to the face (£0·05) and for labour rubbing down the face of the concrete with carborundum stone to produce the finish required (£0·25).

Plywood panels

If plywood panel formwork is specified, take 1 square metre of plywood and *half* the normal timber allowance, everything else remaining the same.

MEMORANDA

Timber in formwork	*Per square metre*
Location	*Volume in m^3*
Walls and foundations	$\frac{1}{16}$
Soffits of floors and roofs	$\frac{1}{12}$
Columns and suspended beams	$\frac{1}{12}$

Divide the foregoing by the assessed number of uses

Labour outputs
Fabrication of formwork—36 carpenter hours per m^3

Fixing and striking	*Per square metre* *Labour hours*
Walls and foundations	$1\frac{1}{2}$
Soffits of floors and roofs	$1\frac{1}{2}$
Columns and suspended beams	$2\frac{1}{2}$

Analyses

Formwork

1 *Sawn formwork to vertical faces of walls*   Per square metre
   Assuming 10 uses and timber @ £28·00 per m³ delivered site

   Materials   £
     Timber 1 m³ @ £28·00 m³   = 28·00
     Unload 2 hours @ £0·64   = 1·28
                                                    = 29·28
     Waste in conversion 5%   = 1·46
     Fabricate: Carpenter 36 hours @ £0·74   = 26·64
                                       per m³ = 57·38

                                       £   £
     Timber $\frac{1}{16}$ m³ @ £57·38   = 3·59
       ÷ 10 uses   = 0·36
     Waste and maintenance 10%   = 0·04
     Fixing and striking 1½ hours @ £0·67   = 1·01
     Mould oil, bolts, wedges   = 0·25
                                      per m² = 1·66

NOTE

Observe the two separate allowances for waste—the first in fabricating the formwork, the second during its use on site.

2 *Sawn formwork to suspended floors*   Per square metre
   Assume 8 uses   £   £
   Timber $\frac{1}{12}$ m³ @ £57·38 m³ as item 1 = 4·78
       ÷ 8 uses   = 0·60
   Waste and maintenance 10%   = 0·06
   Fixing and striking 1½ hours @ £0·67   = 1·01
   Mould oil and telescopic struts   = 0·25
                               per m² = 1·92

3 *Wrought formwork to columns*   Per square metre
   Assume 6 uses   £   £
   Timber $\frac{1}{12}$ m³ @ £57·38 m³ as item 1 = 4·78
       ÷ 6 uses   = 0·80
   Waste and maintenance 10%   = 0·08
   Fixing and striking 2½ hours @ £0·67   = 1·68
   Mould oil and column cramps   = 0·25
                                          = 2·81
   Extra for wrought formwork   = 0·30
                               per m² = 3·11

CONCRETE WORK 41

NOTES

a   Observe the effect on the cost of varying the number of uses.
b   The build-up for formwork to suspended beams is exactly as formwork to columns with the exception that telescopic struts take the place of column cramps.
c   Note that mould oil is included in *every* formwork build-up.

5 Brickwork and Blockwork

The Unit of Measurement is the Metre

Metric brick size

The metric brick is 215 mm long × 102·5 mm wide × 65 mm deep. Mortar joints are 10 mm wide.
There are 60 bricks in one square metre of half brick wall.
There are 120 bricks in one square metre of one brick wall.

Unloading

Allow 2 labourer hours per thousand bricks for unloading and stacking commons or facings.

Waste

It is not customary to allow for waste on common bricks since half-bats and broken bricks can usually be built in as part of the bond. With facing bricks, however, it may be necessary to make some provision if the bricks are soft or brittle or if the nature of the work appears to demand more than the average amount of cutting. In these circumstances an allowance of between 2% and 5% for waste should be made.

Combined labour rate

		£
1 bricklayer		= 0·74
½ labourer @ £0·64		= 0·32
Combined bricklayer rate		= 1·06 per hour

Labour outputs—common bricks

Wall thickness		Bricks per hour
HB	102·5 mm	50
1B	215 mm	60
1½B	327·5 mm	70
Reduced brickwork		80

Multipliers on brickwork

Brickwork	ADD		
in arches	ADD	extra labour	400%
in panels		do	100%
in underpinning		do	100%
to curves		do	50%
in manholes		do	33%
to batter		do	25%
built overhand		do	25%

BRICKWORK AND BLOCKWORK 43

Scaffolding and hoists

Scaffolding and hoists for the complete contract are priced in the Preliminaries.

Mortar

Mixes

The usual mortar mixes are cement–mortar 1:3 or 1:4 and cement–lime–mortar 1:1:6 or 1:2:9.

Analyses

1   *Cement mortar 1:3*   *Per cubic metre*
                                                            £
    Portland cement per tonne   =   7·90
    Unload   1 hour @ £0·64   =   0·64
                                       per tonne =   8·54
                                                            £
    1 m³ Portland cement = 1·5 tonnes @ £8·54   =   12·81
    3 m³ sand @ £2·40 m³   =   7·20
                                                    =   20·01
        add $33\frac{1}{3}\%$ for shrinkage   =   6·67
                                                    =   26·68
        ÷ 4 parts   =   6·67
    Mixing   £
    $5/3\frac{1}{2}$ mortar mixer per hour   = 0·18
    Mixer driver   = 0·64
    NWR plus rate   = 0·02
                                              = 0·84
    Output 1 m³ per hour   =   0·84
                                              per m³ =   7·51

2 *Cement–lime–mortar 1:1:6* *Per cubic metre*

		£
Hydrated lime per tonne	=	8·25
Unload 1 hour @ £0·64	=	0·64
	=	8·89
1 m³ hydrated lime = 0·75 tonnes	=	6·67

		£
1 m³ Portland cement	=	12·81
1 m³ hydrated lime	=	6·67
6 m³ sand @ £2·40 m³	=	14·40
	=	33·88
add 33⅓% for shrinkage	=	11·29
	=	45·17
÷ 8 parts	=	5·65
Mixing as before	=	0·84
per m³ =		6·49

NOTES

a Hydrated lime weighs half as much as Portland cement volume for volume.
b Shrinkage allowance is 33⅓% for mortars (50% for concrete).

Hand mixing

Allow 4 hours per m³ for hand mixing mortar.
  Cement–mortar 1:3 *Per cubic metre*

		£
Materials as above	=	6·67
Hand mixing 4 hours @ £0·64	=	2·56
per m³ =		9·23

  Cement–line–mortar 1:1:6 *Per cubic metre*

		£
Materials as above	=	5·65
Hand mixing 4 hours @ £0·64	=	2·56
per m³ =		8·21

Volume of mortar

The volume of mortar required for 1 square metre of brickwork, per half brick thickness of wall is $\frac{1}{30}$ m³.

BRICKWORK AND BLOCKWORK   45

Analyses

1 *One brick wall in common bricks in*   *Per square metre*
   cement–mortar 1:3   £
   Common bricks @ £9·75 per thousand   =   9·75
   Unload   2 hours @ £0·64   =   1·28
                                 per thousand = 11·03
                                                                    £
   120 bricks @ £11·03 per thousand   =   1·32
   Cement mortar 1:3 $\frac{1}{15}$ m³ @ £7·51 m³   =   0·50
   Labour   @ 60 bricks per hour
            = 2 hours @ £1·06   =   2·12
                                     per m² =   3·94

2 *Half brick wall in common bricks in*   *Per square metre*
   cement mortar 1:3   £
   60 bricks @ £11·03 per thousand   =   0·66
   Cement mortar 1:3 $\frac{1}{30}$ m³ @ £7·51 m³   =   0·25
   Labour   @ 50 bricks per hour
            = 1$\frac{1}{5}$ hours @ £1·06   =   ·1·27
                                     per m² =   2·18

Note the cost effect of the reduced output taken for the half brick wall compared to the one brick wall.

3 *Form cavity between two half brick walls*   *Per square metre*
   including two twisted wall ties per square metre   £
   Wall ties   2 @ £0·01 each   =   0·02
   Labour $\frac{1}{10}$ hour @ £1·06   =   0·11
                                     per m² =   0·13

Facings

MEMORANDA

1 Facing bricks are measured as 'extra over' the cost of common brickwork. The number of facing bricks per square metre varies according to the bond used, as follows:

Bond	No. of facings per m²
Stretcher	60
Flemish	80
English	90

2 The rate of laying facings is less than that of common bricks because of the greater care required and is taken in the examples as 40 facings per hour.

Analyses

Facing bricks

1 Extra over common brickwork for facings PC
   £20 per thousand in English bond and flush
   pointing as the work proceeds   *Per square metre*

	£	
Facings	= 20·00	
Commons	= 9·75	
Extra cost of facings per thousand	= 10·25	

		£
90 facing bricks @ extra £10·25 per thousand		= 0·92
Waste $2\frac{1}{2}\%$		= 0·02
Extra labour: £		
Facings $\frac{90}{40}$ @ £1·06 = 2·39		
Commons $\frac{90}{60}$ @ £1·06 = 1·59		
Extra = 0·80	= 0·80	
Pointing $\frac{1}{4}$ hour @ £1·06	= 0·27	
Mortar in pointing nil	—	
per m² =	2·01	

NOTE

The cost of unloading facings is assumed to be the same as commons so no adjustment is required for unloading.

2 *Ditto in Flemish bond*   *Per square metre*
                                                                £

80 facing bricks @ extra £10·25 per thousand		= 0·82
Waste $2\frac{1}{2}\%$		= 0·02
Extra labour: £		
Facings $\frac{80}{40}$ @ £1·06 = 2·12		
Commons $\frac{80}{60}$ @ £1·06 = 1·41		
Extra = 0·71	= 0·71	
Pointing as item 1	= 0·27	
per m² =	1·82	

NOTE

Walls built entirely of facing bricks are not measured as 'extra over' common brickwork but as facing brick walls and are priced at the full rate, as follows:

BRICKWORK AND BLOCKWORK

3 *Half brick wall in facing bricks PC £25·50 per thousand in cement mortar 1:3 in stretcher bond and raking out joints and weather pointing in coloured cement mortar* *Per square metre*

	£	
Facing bricks per thousand	= 25·50	
Unload 2 hours @ £0·64	= 1·28	
	= 26·78	£
60 facing bricks @ £26·78 per thousand		= 1·61
Waste $2\frac{1}{2}\%$		= 0·04
Cement mortar $\frac{1}{30}$ m³ @ £7·51 m³		= 0·25
Labour @ 40 bricks per hour		
= $1\frac{1}{2}$ hours @ £1·06		= 1·59
Pointing $\frac{3}{4}$ hour @ £1·06		= 0·80
Pointing mortar $\frac{1}{200}$ m³ @ £15·00 m³		= 0·08
	per m² =	4·37

NOTES

a Observe the longer time required for raking out joints and pointing afterwards in coloured cement compared with pointing as the work proceeds.
b Note the allowance for the coloured pointing mortar when the joints are raked out and pointed afterwards.

4 *One brick wall in facing bricks PC £25·50 per thousand in cement mortar 1:3 in Flemish bond including flush pointing both sides as the work proceeds* *Per square metre*

	£
120 facing bricks @ £26·78 per thousand	= 3·21
Waste $2\frac{1}{2}\%$	= 0·08
Cement mortar $\frac{1}{15}$ m³ @ £7·51 m³	= 0·50
Labour @ 40 bricks per hour	
= 3 hours @ £1·06	= 3·18
Pointing $\frac{1}{4}$ hour @ £1·06 = £0·27 × 2 sides	= 0·54
per m² =	7·51

5 *Half brick wall in facing bricks PC £25·50 per thousand in cement mortar 1:3 in Flemish bond with snapped headers as outer skin of cavity wall, including flush pointing.*   Per square metre

	£
60 facing bricks @ £26·78 per thousand	= 1·61
Waste 2½%	= 0·04
Cement mortar $\frac{1}{30}$ m³ @ £7·51 m³	= 0·25
Labour at 40 bricks per hour 80 bricks or half-bricks = 2 hours @ £1·06	= 2·12
Labour snapping 20 bricks ¼ hour @ £1·06	= 0·27
Waste in snapping 2 bricks @ £0·03 each	= 0·06
Pointing ¼ hour @ £1·06	= 0·27
per m² =	__4·62__

NOTES

a One square metre of half brick wall requires 60 bricks.
One square metre of Flemish bond facing requires 80 bricks, i.e. 40 stretchers and 40 headers.
In this example we require 40 stretchers plus 20 bricks cut in half, or 'snapped' to provide 40 headers.
Therefore whilst only 60 whole bricks are required, the bricklayer has 80 bricks, or snapped bricks to lay. This accounts for the labour laying being 2 hours, i.e. 80 at 40 per hour. It is assumed that snapped bricks take the same time to lay as whole bricks.

b Note also the additional waste involved in the snapping process.

Damp proof courses

Damp proof courses over 225 mm or one brick wide are measured in square metres; if less than 225 mm wide they are measured in linear metres stating the width.

In pricing an item of dpc measured in linear metres, build up the price in square metres and convert to linear metres at the end, as shown in the example.

Waste and laps on dpc—5%.
Labour output for 1 bricklayer and ½ labourer—5 m² per hour.

BRICKWORK AND BLOCKWORK 49

Analyses

1 *Bituminous felt damp proof course 102·5 mm*   Per linear metre
   wide and laying on brick walls   £
   Bituminous felt dpc £0·25 per m²   = 0·25
   Waste 5%   = 0·01
   Labour ⅕ hour @ £1·06   = 0·21
                                      per m² = $\overline{0\cdot 47}$
           × 102·5 mm wide   per m = $\underline{0\cdot 05}$

Block partitions

MEMORANDA	*Per square metre*			
	Unload		*Mortar*	*Output*
Thickness	*labour hour*	*Waste*	*m³*	*hours/m²*
75 mm	$\frac{1}{25}$	5%	$\frac{1}{150}$	$\frac{1}{2}$
100 mm	$\frac{1}{20}$	5%	$\frac{1}{100}$	$\frac{2}{3}$
150 mm	$\frac{1}{18}$	5%	$\frac{1}{75}$	$\frac{3}{4}$
225 mm	$\frac{1}{15}$	5%	$\frac{1}{50}$	1

NOTES

a The above table applies to all types of block partitions, solid or hollow, concrete, clinker, ballast, breeze or foamed slag.
b The labour rate is 1 bricklayer and ½ labourer, as in brickwork.
c Fair face one side as the work proceeds—⅛ hour per square metre.
d Raking cutting and waste is measured in linear metres. Allow ⅙ m² of blockwork for waste at the supply rate and ⅙ hour for labour cutting per linear metre.
e Curved cutting and waste—add 50% to the raking cutting price.

Analyses

Block partitions

1   *100 mm concrete block partition in*   *Per square metre*
    cement mortar 1:3   £
    100 mm concrete blocks £0·70 per m^2 = 0·70
    Unload $\frac{1}{20}$ hour @ £0·64   = 0·03
                                                 = 0·73
    Waste   5%   = 0·04
    Cement mortar $\frac{1}{100}$ m^3 @ £7·51 m^3 = 0·08
    Labour $\frac{2}{3}$ hour @ £1·06   = 0·71
                                         per m^2 = 1·56

2   *Raking cutting and waste on 100 mm block*   *Per linear metre*
    partition   £
    Waste $\frac{1}{6}$ m^2 blocks @ £0·73 m^2 = 0·12
    Labour $\frac{1}{6}$ hour @ £1·06   = 0·18
                                         per m = 0·30

6 Rubble Walling

The Unit of Measurement is the Metre

MEMORANDA

A cubic metre of rubble walling requires 2 tonnes of stone and $\frac{1}{4}$ m³ of mortar. One mason and one labourer take 5 hours to lay a cubic metre of rough rubble walling.

Analyses

1 *Random rubble in walls 450 mm thick in lime mortar 1:3*

Per square metre
£

The rate is built up as a price per cubic metre and converted at the end.

2 tonnes rubble stone @ £5·75 per tonne	= 11·50
Waste included	—
Lime mortar $\frac{1}{4}$ m³ @ £5·46 m³	= 1·37
Mason 5 hours @ £0·74	= 3·70
Labourer 5 hours @ £0·64	= 3·20
per m³ =	19·77

450 mm thick = $\frac{450}{1000}$ = per m² = 8·76

7 Roofing

The Unit of Measurement is the Metre

Slate roofing

MEMORANDA

The formulae for calculating the gauge of slating are as follows:
Centre nailed slates:
$$\frac{\text{Length of slate} - \text{lap}}{2}$$
Head nailed slates:
$$\frac{(\text{Length of slate} - 25\text{ mm}) - \text{lap}}{2}$$

The lap for slating is normally 75 mm.

Number of slates

The number of slates required to cover a square metre is calculated as follows:
$$\frac{1 \text{ m}^2}{\text{Width of slate} \times \text{gauge}}$$

EXAMPLE

510 mm × 255 mm slates centre nailed to a 75 mm lap.

$$\text{Gauge} = \frac{510 \text{ mm} - 75 \text{ mm}}{2}$$

$$= 217 \cdot 5 \text{ mm}$$

$$\text{Number of slates} = \frac{1 \text{ m}^2}{0 \cdot 255 \times 0 \cdot 217}$$

$$= \frac{1}{0 \cdot 055} = \underline{18 \text{ no. per m}^2}$$

MEMORANDA

Unloading	2 to 3 labourer hours per 1000 slates, depending on size
Waste	5% on slates
	10% on battens and nails
Nails	two nails per slate
	38 mm nails—350 per kg
	50 mm nails—200 per kg

ROOFING

Length of battens

To calculate the length of battens per square metre of slating laid to a 75 mm lap: multiply the number of slates per m² by the width of the slates.

EXAMPLE

Slates 255 mm × 205 mm: no. per m² = 54
Length of battens = 54 × 0·205 m
= 11 metres

	Slates and battens	Per square metre	
Size of slates mm	Gauge centre-nailed mm	Battens m	Slates no.
255 × 205	90	11	54
410 × 205	167	6	30
510 × 255	217	5	18
610 × 305	267	4	12

Labour laying slates

The normal practice is 1 labourer serving 2 slaters, or $\frac{1}{2}$ labourer per slater.

Size of slates mm	Slates per hour Slater + $\frac{1}{2}$ labourer
255 × 205	90
410 × 205	70
510 × 255	60
610 × 305	50

Tile roofing

MEMORANDA

Size   Plain tiles are 265 mm × 165 mm in size.
Lap   The lap varies from 65 mm to 100 mm, 65 mm being the normal.
Gauge   For a 65 mm lap, the gauge is 100 mm.
No. of tiles   To a 100 mm gauge: 60 tiles per square metre.
Unloading   1 hour labourer per 1000 tiles.
Waste   5% on tiles
   10% on battens and nails.
Nails   Two nails per tile. 120 nails per square metre if nailed every course.
It is normal practice to nail every fourth course = 30 nails per square metre. Allow $\frac{1}{10}$ kg nails per square metre.
Battens   The calculation is as for slating battens, i.e. 60 × 0·165 m = 10 metres.
Labour laying   1 tiler and $\frac{1}{2}$ labourer
   Battens: 50 metres per hour
   Tiles: 100 no. per hour.

Analyses

Slating and tiling

1   *510 mm × 255 mm slates to roof slopes laid to a 75 mm lap laid on and including 40 mm × 20 mm softwood battens*   Per square metre

	£	
Slates per thousand	= 132·00	
Unload 3 hours @ £0·64	= 1·92	
	= 133·92	£
18 slates @ £133·92 per thousand		= 2·41
Waste 5%		= 0·12
Nails $\frac{1}{5}$ kg @ £0·35 kg		= 0·07
Labour 18 slates @ 60 per hour		
= $\frac{18}{60}$ hour @ £1·06		= 0·32
		= 2·92
Battens 5 m @ £0·02 m		= 0·10
Waste 10%		= 0·01
Nails $\frac{1}{10}$ kg @ £0·10 kg		= 0·01
Labour 5 m @ 50 m per hour		
= $\frac{1}{10}$ hour @ £1·06		= 0·11
	per m² =	3·15

ROOFING

2 *Plain tiling to roof slopes laid to a 100 mm gauge nailed every fourth course laid on and including 40 mm × 20 mm softwood battens* Per square metre

	£	
Plain tiles per thousand	= 18·75	
Unload 1 hour @ £0·64	= 0·64	
	= 19·39	£
60 tiles @ £19·39 per thousand		= 1·16
Waste 5%		= 0·06
Nails $\frac{1}{10}$ kg @ £0·30 kg		= 0·03
Labour 60 tiles @ 100 per hour		
= $\frac{3}{5}$ hour @ £1·06		= 0·64
Battens 10 m @ £0·02 m		= 0·20
Waste 10%		= 0·02
Nails $\frac{1}{5}$ kg @ £0·10 kg		= 0·02
Labour 10 m @ 50 m per hour		
= $\frac{1}{5}$ hour @ £1·06		= 0·21
	per m^2 =	2·34

Felt roofing

MEMORANDA

Roofing felt is supplied in rolls 10 metres long by 1 metre wide.
Waste   Allow 10% for waste and laps.
Nails $\frac{1}{10}$ kg per square metre.
Labour $\frac{1}{12}$ tradesman hour per m^2 per layer of felt.

Analyses

1 *One layer bitumen roofing felt and fixing on boarded surfaces*   Per square metre

	£	£
Roofing felt £1·40 per 10 m roll	= 0·14	
Unload	= 0·01	= 0·15
Waste and laps 10%		= 0·02
Labour $\frac{1}{12}$ hour @ £0·74		= 0·06
Nails $\frac{1}{10}$ kg @ £0·20 kg		= 0·02
	per m^2 =	0·25

2 *Raking cutting*   Per linear metre

	£
Waste $\frac{1}{6}$ m^2 @ £0·15 m^2	= 0·03
Labour $\frac{1}{25}$ hour @ £0·74	= 0·03
per m =	0·06

Sheet metal roofing

MEMORANDA

Sheet lead is sold by the tonne, in varying thicknesses weighing 20, 25 or 30 kg per square metre.
Unload   1 hour per tonne.
Waste   1%.
Labour   Plumber and mate laying 25 kg lead to flat roofs, 2 hours per m^2.
             For flashings, valleys and aprons, 3 hours per m^2.
             Stepped flashings, $3\frac{1}{2}$ hours per m^2.

Copper, zinc and aluminium

Waste   2%.
Labour   Add 20% extra labour over sheet lead.

Analyses

1 *Sheet lead 25 kg per square metre and laying on flat roofs*   *Per square metre*
                                                                      £
Sheet lead @ £190·00 per tonne   = 190·00
Unload   1 hour @ £0·64   =   0·64
                                                                 = 190·64
                                                                     £

$$1\ m^2 = 25\ kg = \frac{25}{1000}\qquad = \quad 4\cdot77$$

Waste   1%   =   0·05
Labour   2 hours @ £1·38   =   2·76
                                                      per m^2 =   7·58

2 *Ditto in stepped flashings 230 mm girth*   *Per linear metre*
                                                               £
Sheet lead as above   = 4·77
Waste   1%   = 0·05
Labour $3\frac{1}{2}$ hours @ £1·38   = 4·83
                                                                  = 9·65

$$230\ mm\ girth = \frac{230}{1000} = \qquad per\ m = 2\cdot22$$

8 Carpentry

The Unit of Measurement is the Metre.

MEMORANDA

1. The basic unit of supply for carcassing timber used in carpentry is the cubic metre. Structural timber is measured in linear metres stating the size.
2. Unloading   Allow 2 hours per cubic metre.
3. Waste   10%.
4. Labour   To provide for labourer attendance to carpenters, allow $\frac{1}{8}$ labourer per carpenter. The joint labour rate is therefore:

	£
Carpenter	0·74
$\frac{1}{8}$ labourer @ £0·64	0·08
Rate per carpenter/hour	0·82

5. Nails   Allow 3 kg per cubic metre.

Labour outputs—carcassing timber

Item	Carpenter hours per m^3
Wall plates	18
Floor and ceiling joists	24
Rafters	36
Purlins, ties, struts and partitions	48
Roof trusses	60

Roof boarding

To flat or sloping roofs   *Per square metre*

Unload $\frac{1}{30}$ hour per m^2
Waste   5%.
Labour $\frac{1}{2}$ hour for plain edged boarding
    $\frac{2}{3}$ hour for tongued and grooved boarding
Nails $\frac{1}{10}$ kg
Raking cutting   Waste: $\frac{1}{6}$ m^2
    Labour: $\frac{1}{4}$ hour per linear metre
Curved cutting   Add 50% to raking cutting rate

Holes in timber	*Carpenter hours per hole*
Bore 25 mm timber for 12 mm bolt	$\frac{1}{12}$
Ditto 75 mm ditto	$\frac{1}{8}$
Ditto 150 mm ditto	$\frac{1}{4}$
Extra for letting in head of bolt flush	$\frac{1}{12}$

Wall boarding — *Per square metre*
Unload $\frac{1}{50}$ hour per m^2
Waste 5%
Nails $\frac{1}{10}$ kg

Labour outputs

Item	Carpenter hours per m^2
12 mm Insulation boarding laid with butt joints on joists	$\frac{1}{5}$
12 mm Ditto fixed vertically to stud partition	$\frac{1}{4}$
25 mm Fibreglass laid on joists	$\frac{1}{6}$
6 mm Asbestos cement sheeting fixed to soffit of joists	$\frac{1}{3}$
6 mm Ditto fixed vertically to stud partition	$\frac{1}{4}$
50 mm Wood wool slabs laid on joists	$\frac{1}{4}$

Analyses

1 *Carcassing timber in floor joists 200 mm × 50 mm*   *Per linear metre*

£
Timber   £28·00 per m^3 = 28·00
Unload   2 hours @ £0·64   = 1·28
                                     = 29·28
Waste   10%   = 2·93
Labour   24 hours @ £0·82   = 19·68
Nails   3 kg @ £0·10 kg   = 0·30
                           per m^3 = 52·19

1 linear metre 200 mm × 50 mm
$= 1 \times \frac{200}{1000} \times \frac{50}{1000} = \frac{1}{100}$ m^3

per m = £0·52

2 *Ditto in purlins 100 mm × 150 mm*   *Per linear metre*

£
Timber as above   = 28·00
Unload   ditto   = 1·28
Waste   10%   = 2·93
Labour   48 hours @ £0·82 = 39·36
Nails   3 kg @ £0·10 kg   = 0·30
                 per m^3 = 71·87

$1 \times \frac{100}{1000} \times \frac{150}{1000} =$

per m = £1·08

CARPENTRY

NOTE

Rates for carcassing timber are best built up on a cubic metre basis then converted to linear metres of the appropriate dimensions.

3 *20 mm plain edged roof boarding*   *Per square metre*
    £

20 mm roof boarding @ £0·60 per m²   = 0·60
  Unload $\frac{1}{30}$ hour @ £0·64   = 0·02
  Waste   5%   = 0·03
  Labour $\frac{1}{2}$ hour @ £0·82   = 0·41
  Nails $\frac{1}{10}$ kg @ £0·10 kg   = 0·01
    per m² = $\underline{1·07}$

4 *Raking cutting*   *Per linear metre*
    £

  Waste $\frac{1}{6}$ m² @ £0·62 m²   = 0·10
  Labour $\frac{1}{4}$ hour @ £0·82   = 0·21
    per m = $\underline{0·31}$

5 *Curved cutting*   *Per linear metre*
    £

  As item 4   = 0·31
  Add 50%   = 0·16
    per m = $\underline{0·47}$

6 *12 mm insulation boarding fixed vertically*   *Per square metre*
  to stud partition   £
12 mm insulation boarding @ £0·30 per m²   = 0·30
  Unload $\frac{1}{50}$ hour @ £0·64   = 0·01
  Waste   5%   = 0·02
  Labour $\frac{1}{4}$ hour @ £0·82   = 0·21
  Nails $\frac{1}{10}$ kg @ £0·10 kg   = 0·01
    per m² = $\underline{0·55}$

9 Joinery

The Unit of Measurement is the Metre

MEMORANDA

1. The combined rate for fixing joinery is the same as that for carpentry—1 joiner with $\frac{1}{8}$ labourer in attendance.
2. The outputs shown are for softwood.
   For hardwood extra labour of from 50% to 150% is added, depending on the relative hardness and difficulty of working the timber.
3. The outputs include for nailing. For screwing add 25% extra labour.

Floor boarding   *Per square metre*

Unload $\frac{1}{30}$ hour per m^2
Waste   5%
Labour $\frac{1}{2}$ hour for plain edged boarding
          $\frac{2}{3}$ hour for tongued and grooved boarding
Nails $\frac{1}{10}$ kg

NOTE

The foregoing outputs are exactly as for roof boarding.

Skirtings, stops, bearers, fillets, mouldings and architraves   *Per linear metre*
Waste   10%

Sectional area	Fixing Joiner hours
Up to 2000 mm^2	$\frac{1}{7}$
2000 to 4000 mm^2	$\frac{1}{5}$
Over 4000 mm^2	$\frac{1}{4}$
Grounds for skirtings	$\frac{1}{15}$

Mitres, fitted ends and fair ends: $\frac{1}{6}$ metre waste and $\frac{1}{10}$ hour joiner
Returned or irregular mitres: $\frac{1}{6}$ metre waste and $\frac{1}{5}$ hour joiner

Doors

Hanging doors on butt hinges   *Each*

Size	Joiner hours
up to 1 m^2	$1\frac{1}{2}$
1 to 2 m^2	2
over 2 m^2	$2\frac{1}{2}$

The foregoing times apply to doors up to 50 mm thick. For doors over 50 mm thick, add 25% extra labour.

JOINERY 61

Door frames and linings *Per linear metre*

Sectional area	Joiner hours
Up to 4000 mm^2	$\frac{1}{15}$
400 to 8000 mm^2	$\frac{1}{12}$
Over 8000 mm^2	$\frac{1}{10}$

Softwood windows *Each*

Size	Unloading Hours	Fixing Joiner hours
Up to $\frac{1}{2}$ m^2	$\frac{1}{20}$	1
$\frac{1}{2}$ to 1 m^2	$\frac{1}{16}$	$1\frac{1}{2}$
1 to 2 m^2	$\frac{1}{12}$	2
2 to 3 m^2	$\frac{1}{8}$	$2\frac{1}{2}$

Plugging *Per linear metre*

Plugging brickwork for joinery: 2 plugs and $\frac{1}{5}$ hour joiner
Plugging concrete: 2 plugs and $\frac{1}{4}$ hour joiner

Analyses

1 *25 mm Tongued and grooved floor boarding* *Per square metre*
    £
    25 mm T & G floor boarding @ £0·80 per m^2 = 0·80
    Unload $\frac{1}{30}$ hour @ £0·64   = 0·02
    Waste 5%   = 0·04
    Labour $\frac{2}{3}$ hour @ £0·82   = 0·55
    Nails $\frac{1}{10}$ kg @ £0·10 kg   = 0·01
                                          per m^2 = $\underline{1·42}$

2 *25 mm Ditto in small quantities* *Per square metre*
    £
    As item 1   = 1·42
    Add 25% extra material = 25% × £0·83   = 0·21
    Add 100% extra labour   = 0·55
                                          per m^2 = $\underline{2·18}$

NOTE

This is the general approach in pricing any item measured in small quantities. The additional waste and extra labour over normal outputs are added to the basic rate.

3 *25 mm × 100 mm softwood skirting plugged and screwed to brickwork*

		Per linear metre £
25 mm × 100 mm softwood skirting @ £0·08 per m		= 0·08
Unload included in output		
Waste 10%		= 0·01
Labour ⅕ hour @ £0·82		= 0·16
Extra labour screwing 25% of £0·16		= 0·04
Screws		= 0·01
Plugging 2 plugs		= 0·01
⅕ hour @ £0·82		= 0·16
	per m =	<u>0·47</u>

4 *50 mm × 100 mm softwood door frame*

		Per linear metre £
50 mm × 100 mm softwood frame @ £0·20 per m		= 0·20
Unload		
Waste 5%		= 0·01
Labour 1/12 hour @ £0·82		= 0·07
Nails		= 0·01
	per m =	<u>0·29</u>

5 *Softwood casement window 1 m × 1·50 m and fixing to brick walls*

		Each £
Window £6·75 each		= 6·75
Unload 1/12 hour @ £0·64		= 0·05
Labour 2 hours @ £0·82		= 1·64
Screws and plugs		= 0·05
	Each =	<u>8·49</u>

JOINERY 63

Ironmongery

MEMORANDA

1 The cost of the necessary screws is normally included in the supply price of the ironmongery.
2 The gross labour rate is as for fixing joinery.
3 The times shown are for fixing to softwood.
   For fixing to hardwood add from 50% to 150% extra labour according to relative hardness of the timber.

Labour fixing

Item	*Joiner hours*
Hat and coat hook	$\frac{1}{4}$
Pair of butt hinges	$\frac{1}{4}$
Pull handle	$\frac{1}{4}$
Push plate	$\frac{1}{4}$
Shelf bracket	$\frac{1}{4}$
Cabin hook and eye	$\frac{1}{4}$
Bales catch	$\frac{1}{3}$
150 mm barrel bolt	$\frac{1}{3}$
Norfolk latch	$\frac{1}{2}$
150 mm flush bolt	$\frac{2}{3}$
Panic bolt	$\frac{3}{4}$
Rim dead lock or latch	$\frac{3}{4}$
Mortice dead lock	1
Cylinder lock	1
WC indicator bolt	1
Mortice lock and furniture	$1\frac{1}{4}$
Rebated mortice dead lock	$1\frac{1}{2}$
Overhead door closer	$1\frac{1}{2}$
Postal plate	$1\frac{1}{2}$

10 Structural Steelwork and Metalwork

The Unit of Measurement is the Kilogramme

MEMORANDA

1. Structural steelwork comprising rolled steel joists, channels and tees is quoted on the basis of a price per tonne.
2. Unloading   2 hours per tonne.
3. Waste and rolling margin   5%.
4. Hoisting and fixing. For small quantities allow 60 hours per tonne for joists and channel sections and 80 hours per tonne for angles and tees.
   Balustrades—160 to 200 hours per tonne.
   For a complete steel framed building, a crane would be utilised and priced for the number of weeks required on site.

	Fixing bolts	*Each*
Size		*Carpenter hours*
Up to 150 mm long		$\frac{1}{12}$
150 mm to 300 mm long		$\frac{1}{6}$

	Metal windows	*Each*
	Unloading	*Fixing*
Size	*hours*	*Joiner hours*
Up to $\frac{1}{2}$ m²	$\frac{1}{20}$	1
$\frac{1}{2}$ to 1 m²	$\frac{1}{16}$	$1\frac{1}{2}$
1 to 2 m²	$\frac{1}{12}$	2
2 to 3 m²	$\frac{1}{8}$	$2\frac{1}{2}$

11 Plumbing

The Unit of Measurement is the Metre

MEMORANDA

Unloading   Unloading and handling of materials is carried out by the plumber's mate. No specific unloading is therefore shown.
Waste   On all piping and fittings—5%.
Labour   The joint rate for a plumber and mate is used for all fixing:

	£
Plumber	0·74
Mate	0·64
Per hour	1·38

Labour outputs

Piping	*Per linear metre*
	Plumber and mate
	Hours
Rainwater pipes	
75 mm cast iron half round gutter fixed to fascia	$\frac{1}{3}$
75 mm cast iron rainwater pipe plugged to brickwork	$\frac{1}{3}$
Lead piping	
13 mm lead piping fixed to brick walls	$\frac{1}{4}$
50 mm ditto	$\frac{1}{3}$
Galvanised or copper piping	
13 mm piping fixed to walls	$\frac{1}{4}$
50 mm ditto	$\frac{1}{2}$

NOTES

a   For intermediate pipe sizes use pro-rata times.
b   For piping laid in trenches allow half the foregoing labour.

Lead piping—wiped soldered joints

		Solder kg	Each Hours
Soldered joint to	13 mm pipe	$\frac{1}{4}$	$\frac{1}{2}$
ditto	25 mm pipe	$\frac{1}{2}$	1
ditto	100 mm pipe	2	2

NOTES

a   For running joints allow half the above material and labour.
b   For intermediate sizes use pro-rata times.

Fittings	Each
Labour fixing	Hours
Lead P trap	1
Bibcock or stopcock	½

Sanitary fittings	Each
Labour fixing	Hours
Bath	2
Lavatory basin	1½
WC	2½
Sink	1½
Sink brackets	1½
Bath fittings	1
Lavatory basin or sink fittings	1

Analyses

1   *75 mm diameter plastic half-round rainwater gutter fixed on fascia brackets*   *Per linear metre*
   £
   75 mm gutter @ £0·25 per m   = 0·25
   Waste 5%   = 0·01
   Plumber and mate ¼ hour @ £1·38   = 0·35
   Fascia bracket 1 per m   = 0·09
   Screws   = 0·01
   per m = <u>0·71</u>

2   *75 mm diameter cast iron rainwater pipe jointed in red lead and plugged to brick walls*   *Per linear metre*
   £
   75 mm rainwater pipe @ £0·70 per m = 0·70
   Waste 5%   = 0·04
   Plumber and mate ⅓ hour @ £1·38   = 0·46
   Red lead and yarn   = 0·02
   Pipe nails and plugs   = 0·01
   per m = <u>1·23</u>

PLUMBING

3 *20 mm lead piping including running joints*   *Per linear metre*
   and fixing to brick walls   £
   20 mm lead pipe @ £0·48 per m   = 0·48
   Waste 5%   = 0·02
   Plumber and mate $\frac{1}{4}$ hour @ £1·38   = 0·35
   Running joints: £
      $\frac{1}{6}$ kg solder @ £0·74 kg   = 0·12
      $\frac{1}{3}$ hour @ £1·38   = 0·46
      per joint = 0·58
     1 joint to 4 linear metres = ÷4   = 0·15
   Pipe clips and screws   = 0·05
      per m = 1·05

4 *20 mm ditto and laying in trench*   *Per linear metre*
        £
   20 mm lead pipe and waste as above   = 0·50
   Plumber and mate $\frac{1}{8}$ hour @ £1·38   = 0·17
   Running joints 1 joint to 6 linear metres
       = £0·58 ÷ 6   = 0·10
      per m = 0·77

5 *75 mm wiped soldered joint*   *Each*
        £
   $1\frac{1}{2}$ kg solder @ £0·74 kg   = 1·11
   Plumber and mate $1\frac{1}{2}$ hours @ £1·38   = 2·07
      each = 3·18

6 *25 mm diameter copper piping and*   *Per linear metre*
   fixing to brick walls   £
   25 mm copper pipe @ £0·74 per m   = 0·74
   Waste 5%   = 0·04
   Plumber and mate $\frac{1}{3}$ hour @ £1·38   = 0·46
   Pipe clip 1 @ £0·02 each   = 0·02
   Screws and plugs 2 @ £0·01 each   = 0·02
      per m = 1·28

7 *Fix only white glazed lavatory basin*   *Each*
   including fittings   £
   Plumber and mate
      Lavatory basin $1\frac{1}{2}$
      Fittings   1
          $2\frac{1}{2}$
   $2\frac{1}{2}$ hours @ £1·38   = 3·45
   Sundry materials red lead, washers, hemp,
          plugs and screws   = 0·20
      each = 3·65

12 Plasterwork and Finishings

The Unit of Measurement is the Metre

MEMORANDA

Weight of materials

	Tonnes per m^3
Portland cement	1·5
Sand	1·5
Hydrated lime	0·75
Gypsum plaster	1·0

Plastering mixes
Cement mortar   1:3
Lime mortar   1:3
Hardwall plaster render coat   1:2
Ditto setting coat   Neat plaster

Thickness of plastering

	mm
Render coat	13
Floating coat	10
Setting coat	3
Render and set	16
Render, float and set	26

Hand mixing
Allow 4 labourer hours per m^3
Waste   10%

Combined labour rate
1 plasterer and $\frac{1}{2}$ labourer = £1·06 per hour

Labour outputs	Per square metre
Plastering walls and soffits	Hours
Rough render	$\frac{1}{4}$
Plain face trowelled smooth	$\frac{1}{3}$
Render and set	$\frac{1}{2}$
Render, float and set	$\frac{2}{3}$
Skim coat	$\frac{1}{4}$

Plastering in narrow widths	Per linear metre
	Extra labour
Uo tp 100 mm wide	75%
100 mm to 200 mm wide	50%
200 mm to 300 mm wide	25%

PLASTERWORK AND FINISHINGS   69

Plasterboard   *Per square metre*
Unload $\frac{1}{50}$ hour per m^2
Waste   5%
Nails $\frac{1}{10}$ kg

Labour outputs   *Hours*
10 mm plasterboard fixed vertically to stud partitions $\frac{1}{5}$
10 mm ditto fixed to soffit of joists $\frac{1}{4}$

Analyses

1 *13 mm rough render in cemenet mortar 1:3*   *Per square metre*
   Cement mortar 1:3   £
      Portland cement per tonne   = 7·90
      Unload   1 hour @ £0·64   = 0·64
                          per tonne = $\overline{8·54}$
                                                            £
   1 m^3 Portland cement = 1·5 tonnes @ £8·54   = 12·81
   3 m^3 sand @ £2·40 m^3 = 7·20
                                                     = 20·01
      Add   33$\frac{1}{3}$% for shrinkage   = 6·67
                                                     = 26·68
      ÷ 4 parts   = 6·67
   Hand mixing   4 hours @ £0·64   = 2·56
                                          per m^3 = 9·23
                                                            £
   13 mm cement mortar 1:3 @ £9·23 m^3 = 0·12
   Waste   10%   = 0·01
   Labour $\frac{1}{4}$ hour @ £1·06   = 0·27
                                          per m^2 = $\underline{0·40}$

PLASTERWORK AND FINISHINGS

2 *Render and set in lime mortar 1:3* *Per square metre*
   Lime mortar 1:3 £
      Hydrated lime per tonne   =   8·25
      Unload   1 hour @ £0·64   =   0·64
                           per tonne =   8·89
                                         £

   1 m³ hydrated lime = 0·75 tonnes
      @ £8·89   =   6·67
   3 m³ sand @ £2·40 m³   =   7·20
                                           = 13·87
   Add $33\frac{1}{3}\%$ for shrinkage   =   4·62
                                           = 18·49
      ÷ 4 parts   =   4·62
   Hand mixing   4 hours @ £0·64   =   2·56
                                per m³ =   7·18   £
   16 mm lime mortar 1:3 @ £7·18 m³   = 0·11
   Waste   10%   = 0·01
   Labour $\frac{1}{2}$ hour @ £1·06   = 0·53
                                                  per m² = <u>0·65</u>

3 *Ditto 200 mm to 300 mm wide*   *Per linear metre*
                                                      £
   As above   m²   = 0·65
   Extra labour   25% × £0·53   = 0·13
                               per m² = 0·78   £

   300 mm wide = $\dfrac{300}{1000}$ =   per m = <u>0·23</u>

   Note that the full width of 300 mm is priced.

4 *Setting cost of hardwall plaster on*   *Per square metre*
   plasterboard   £
   Gypsum plaster per tonne   = 12·90
   Unload   1 hour @ £0·64   =   0·64
                             per tonne = 13·54
   Mixing   4 hours @ £0·64   =   2·56
                                         = 16·10

   1 m³ gypsum plaster = 1 tonne   £
   3 mm gypsum plaster @ £16·10 m³   = 0·05
   Waste   10%   = 0·01
   Labour $\frac{1}{4}$ hour @ £1·06   = 0·27
                                                per m² = <u>0·33</u>

5   *10 mm plasterboard fixed to soffits of joists to*   Per square metre
    receive plaster finish   £
    10 mm plasterboard @ £0·18 per m²   = 0·18
    Unload $\frac{1}{50}$ hour @ £0·64   = 0·01
    Waste   5%   = 0·01
    Labour $\frac{1}{4}$ hour @ £1·06   = 0·27
    Nails $\frac{1}{10}$ kg @ £0·10 kg   = 0·01
                                         per m² = 0·48

Wall tiling

Analyses

1   *150 mm × 150 mm white glazed wall tiles bedded*
    in cement mortar 1:3 and pointed in white   Per square metre
    cement   £
    Wall tiles £1·20 per m²   = 1·20
    Unload $\frac{1}{50}$ hour @ £0·64   = 0·01
    Waste   5%   = 0·06
    Labour   tiler and $\frac{1}{2}$ labourer: 1$\frac{1}{2}$ hours @ £1·06   = 1·59
    Cement mortar $\frac{1}{100}$ m³ @ £9·23 m³   = 0·09
    White cement   = 0·02
                                          per m² = 2·97

NOTE

For wall tiles stuck to walls with adhesive take half the above labour cost.

2   *Raking cutting and waste*   Per linear metre
                                                   £
    Waste $\frac{1}{6}$ m² @ £1·21 m²   = 0·20
    Labour $\frac{1}{3}$ hour @ £1·06   = 0·35
                                          per m = 0·55

Flooring

MEMORANDA

Allow 5% waste on floor coverings.

Labour outputs	Per square metre
Flooring	*Tradesman hours*
Wood block	$1\frac{1}{4}$
Quarry tiles	$1\frac{1}{4}$
Cork tiles	$\frac{3}{4}$
25 mm granolithic (pavior and half labourer)	$\frac{1}{2}$
Linoleum, rubber or vinyl tiles	$\frac{1}{3}$
Ditto sheeting	$\frac{1}{4}$

Paving slabs and kerbs

MEMORANDA

Paving slabs *Per square metre*
Unload $\frac{1}{8}$ hour
Waste   5%
Labour $\frac{1}{3}$ hour pavior and half labourer
Mortar $\frac{1}{40}$ m³

Raking cutting *Per linear metre*
Waste $\frac{1}{6}$ m²
Labour $\frac{1}{6}$ hour

Curved cutting *Per linear metre*
Add 50% to last

Kerbs and channels *Per linear metre*
Unload $\frac{1}{12}$ hour
Waste $2\frac{1}{2}$%
Labour $\frac{1}{5}$ hour pavior and half labourer
Mortar   for bedding and jointing $\frac{1}{200}$ m³
Pavior rate—a pavior receives £0·04 over a labourer's rate.
Combined labour rate:

		£
	Pavior	0·68
	$\frac{1}{2}$ labourer	0·32
		1·00

PLASTERWORK AND FINISHINGS 73

Analyses

1 *50 mm paving slabs bedded and jointed in cement mortar 1:3*

    Per square metre
    £

50 mm paving slabs £0·90 per m²   = 0·90
Unload $\frac{1}{8}$ hour @ £0·64   = 0·08
     = 0·98

Waste   5%   = 0·05
Labour $\frac{1}{3}$ hour @ £1·00   = 0·33
Mortar $\frac{1}{40}$ m³ @ £9·23 m³   = 0·23

per m² = $\underline{1\cdot59}$

2 *Raking cutting*

    Per linear metre
    £

Waste $\frac{1}{6}$ m² @ £0·98 m²   = 0·16
Labour $\frac{1}{6}$ hour @ £1·00   = 0·17

per m = $\underline{0\cdot33}$

3 *125 mm × 250 mm precast concrete kerb bedded and jointed in cement mortar 1:3*

    Per linear metre
    £

125 mm × 250 mm kerb £0·50 per m   = 0·50
Unload $\frac{1}{12}$ hour @ £0·64   = 0·05
     = 0·55

Waste $2\frac{1}{2}$%   = 0·01
Labour $\frac{1}{5}$ hour @ £1·00   = 0·20
Mortar $\frac{1}{200}$ m³ @ £9·23 m³   = 0·05

per m = $\underline{0\cdot81}$

13 Glazing

The Unit of Measurement is the Metre

MEMORANDA

1. Sheet glass for glazing is supplied in various thicknesses from 2 mm to 4 mm.
   Float glass is normally 6 mm thick, obscured glass is either 3 mm or 6 mm.
2. Sizes of glass panes are classified in the Bills of Quantities as follows:
   Up to $0·10$ m^2
   $0·10$ m^2 to $0·50$ m^2
   $0·50$ m^2 to $1·00$ m^2
   Over $1·00$ m^2
3. Unloading—the labour outputs include for unloading.
4. Waste and breakage—5%.

Labour outputs Glass up to 4 mm thick Panes	Per square metre	
	Glazier hours	Putty kg
Up to $0·10$ m^2	1	3
$0·10$ m^2 to $0·50$ m^2	$\frac{3}{4}$	2
$0·50$ m^2 to $1·00$ m^2	$\frac{2}{3}$	1
Over $1·00$ m^2	$\frac{1}{2}$	$\frac{3}{4}$

The outputs apply to either wood or metal casements.
For 6 mm glass—add 25% extra labour.
For fixing with glazing beads and brass cups and screws—add 25% extra labour.

Analyses

1. *4 mm clear sheet glass in panes*
   $0·50$ m^2 to $1·00$ m^2

	Per square metre £
4 mm glass £1·38 per m^2	= 1·38
Waste 5%	= 0·07
Glazier $\frac{2}{3}$ hour @ £0·74	= 0·49
Putty 1 kg @ £0·07 kg	= 0·07
per m^2 =	<u>2·01</u>

GLAZING

2 *6 mm polished plate glass in panes over 1·00 m²*
   fixed with hardwood glazing beads and brass
   cups and screws   *Per square metre*

   NOTE The glazing beads are measured and priced
         separately.

			£
6 mm polished plate glass £3·36 per m²			= 3·36
Waste 5%			= 0·17
Glazier ½ hour @ £0·74			= 0·37
Extra labour	6 mm glass	25%	
Ditto	cups and screws	25%	
		50% £0·37	= 0·19
		per m²	= 4·09

14 Painting and Decorating

The Unit of Measurement is the Metre

MEMORANDA

Distempers and cement paints are supplied in 50 kg drums. Other paints are supplied by the litre.
Waste—5%.
Painter's rate—add £0·02 per hour to the Tradesman's rate to cover the use and waste of brushes and glasspaper.

	Per coat—labour and materials	*Per square metre*
	Paint	Painter hours
Washable distemper	$\frac{1}{4}$ kg distemper	$\frac{1}{10}$
	$\frac{1}{16}$ litre petrifying liquid	
Emulsion paint	$\frac{1}{10}$ litre	$\frac{1}{10}$
Knotting, priming and stopping	$\frac{1}{50}$ litre knotting	$\frac{1}{5}$
	$\frac{1}{25}$ stopping	
	$\frac{1}{10}$ litre primer	
Oil paint		
Undercoat	$\frac{1}{10}$ litre	$\frac{1}{8}$
Finishing coat	$\frac{1}{10}$ litre	$\frac{1}{5}$

NOTES

a   The outputs for distemper and emulsion paint are for painting to plaster surfaces. For brick and concrete surfaces add 25% extra labour and material.

b   The outputs for oil paint apply equally to wood or metal surfaces.

c   Painting in narrow widths   *Per linear metre*
                                              Extra labour

Up to 100 mm wide	75%
100 mm to 200 mm wide	50%
200 mm to 300 mm wide	25%

Analyses

1   *Two coats washable distemper on plastered walls*   *Per square metre*
                                                               £

Washable distemper $\frac{1}{4}$ kg @ £6·50 per 50 kg   = 0·03
Petrifying liquid $\frac{1}{16}$ litre @ £0·14 litre   = 0·01
                                                              ─────
                                                            = 0·04
Waste 5%   —
Painter $\frac{1}{10}$ hour @ £0·76   = 0·08
                                                              ─────
                                              per coat   = 0·12
                                   × 2 coats = per m²   = <u>0·24</u>

76

PAINTING AND DECORATING 77

2 *Two coats emulsion paint on brick walls*   *Per square metre*
                                                        £
   Emulsion paint $\frac{1}{10}$ litre @ £0·58 litre   = 0·06
   Waste 5%   —
   Painter $\frac{1}{10}$ hour @ £0·76   = 0·08
   Extra labour and material on brick walls
      25% × £0·14   = 0·04
                                         per coat = $\overline{0·18}$
                          × 2 coats = per m² = $\underline{\underline{0·36}}$

3 *Knot, prime, stop and paint two undercoats and*   *Per square metre*
   one finishing coat on general wood surfaces   £
   Knotting $\frac{1}{50}$ litre @ £0·45 litre   = 0·01
   Primer $\frac{1}{10}$ litre @ £0·58 litre   = 0·06
   Stopping $\frac{1}{25}$ kg @ £0·10 kg   = 0·01
   Undercoats $\frac{1}{10}$ litre @ £0·75 litre × 2 coats = 0·15
   Finishing coat $\frac{1}{10}$ litre @ £0·80 litre   = 0·08
                                                        = $\overline{0·31}$
   Waste 5%   = 0·02
   Painter   K, P and S $\frac{1}{5}$ hour @ £0·76   = 0·15
              Undercoats $\frac{1}{10}$ hour × 2 coats
               = $\frac{1}{5}$ hour @ £0·76   = 0·15
              Finishing coat $\frac{1}{8}$ hour @ £0·76   = 0·10
                                          per m² = $\underline{\underline{0·73}}$

4 *Ditto not exceeding 100 mm wide*   *Per linear metre*
                                                        £
   As above per m²   = 0·73
   Extra labour   75% on £0·40   = 0·30
                                          per m² = $\overline{1·03}$

   × 100 mm wide = $\dfrac{100}{1000}$ = per m   = $\underline{\underline{0·10}}$

   Note that the full width of 100 mm is priced.

5 *Ditto 200 mm to 300 mm wide*   *Per linear metre*
                                                        £
   As above per m²   = 0·73
   Extra labour   25% on £0·40   = 0·10
                                          per m² = $\overline{0·83}$

   × 300 mm wide = $\dfrac{300}{1000}$ = per m   = $\underline{\underline{0·25}}$

15 Drainage

The Unit of Measurement is the Metre

Excavation for drain trenches

MEMORANDA

The build up of a rate for excavating drain trenches comprises four items:
1 Excavation
2 Return filling and ramming
3 Disposal of surplus excavated material
4 Planking and strutting

Trench widths for pipes up to 300 mm diameter are as follows:

Depth	Width
Up to 1·50 m	700 mm
1·50 m to 3·00 m	800 mm
3·00 m to 4·50 m	1 m

The excavation can be carried out either completely by hand or partly by machine and partly by hand. The actual proportion of machine to hand excavation to be adopted is a decision taken by the Estimator and is related to the actual project being priced.

Assuming a mechanical trencher at £2·00 per hour and an output of 5 m^3 per hour, the cost of machine excavation is

£2·00 ÷ 5 m^3 per hour = £0·40 m^3.

Hand excavation   3 hours @ £0·64 = £1·92 m^3.

Assuming 75% machine to 25% hand excavation:

	£
75% @ £0·40	= 0·30
25% @ £1·92	= 0·48
per m^3 =	0·78

Return filling and ramming

On average $\frac{7}{8}$ of the total volume excavated is returned to the trench, the remaining $\frac{1}{8}$ being carted away.

Labour output—1 hour per cubic metre.

DRAINAGE

Analyses

1 *Excavate trench for 100 mm drain pipe not*   *Per linear metre*
 exceeding 1·50 m deep, average 1·00 m deep   £
 Excavate   1·0 m × 0·7 m × 1·0 m deep
        = 0·7 m³ @ £0·78 m³   = 0·55
 RFR   0·7 m³ @ £0·64 = £0·45 × $\frac{7}{8}$ = 0·39
 Cart away   0·7 m³ @ £0·90 = £0·63 × $\frac{1}{8}$ = 0·08
 Planking and strutting   2 m² @ £0·16   = 0·32
                                    per m = $\underline{1·34}$

2 *Ditto average 1·30 m deep*   *Per linear metre*
                                          £
 Excavate   1·0 m × 0·7 m × 1·3 m deep
        = 0·9 m³ @ £0·78 m³   = 0·70
 RFR   0·9 m³ @ £0·64 = £0·58 × $\frac{7}{8}$ = 0·51
 Cart away   0·9 m³ @ £0·90 = £0·81 × $\frac{1}{8}$ = 0·10
 Planking and strutting   2 m² @ £0·16   = 0·32
                                    per m = $\underline{1·63}$

Drain pipes   *Per linear metre*

MEMORANDA

Salt glazed stoneware drain pipes are quoted in a standard price list with various percentage adjustments for the quality of pipes and the value of the order.

Unloading   Stoneware and concrete pipes:

Diameter (mm)	Hours
100	$\frac{1}{20}$
150	$\frac{1}{15}$
225	$\frac{1}{10}$
300	$\frac{1}{8}$

Waste   5% on all pipes and fittings.
Labour   Laying and jointing stoneware or concrete pipes:

Diameter mm	Pipelayer Hours
100	$\frac{1}{3}$
150	$\frac{2}{5}$
225	$\frac{1}{2}$
300	$\frac{3}{5}$

A pipelayer is paid £0·02 per hour over a labourer's rate = $\underline{£0·66}$.

Analyses

1 *150 mm salt glazed stoneware drain pipe and*   *Per linear metre*
   lay and joint in trench   £

	£
150 mm pipe £0·30 per m	= 0·30
Unload $\frac{1}{15}$ hour @ £0·64	= 0·04
	= 0·34
Waste 5%	= 0·02
Pipelayer $\frac{2}{5}$ hour @ £0·66	= 0·26
Mortar $\frac{1}{200}$ m³ @ £9·23 m³	= 0·05
Yarn	= 0·01
per m =	0·68

2 *Extra for 150 mm bend*   *Each*
                                                      £

	£
150 mm bend £0·40 each	= 0·40
Unload $\frac{1}{30}$ hour @ £0·64	= 0·02
Waste 5%	= 0·02
Pipelayer $\frac{2}{5}$ hour @ £0·66	= 0·26
Mortar	= 0·05
Yarn	= 0·01
	= 0·76
Ddt $\frac{1}{2}$ linear metre 150 mm pipe @ £0·68 m	= 0·34
extra =	0·42

NOTE

Junctions are also measured as extra over ordinary pipe and are priced in a similar manner.

16 Preliminaries

Preliminaries, or site overheads, are the costs of the contractor's site organisation and they are calculated after the pricing of the Bills of Quantities have been completed.

The Standard Method of Measurement sets out the information to be provided in the Bills, including the location of the site, the mode of access, any limitation of working space and the presence of any adjacent or abutting buildings.

The contractor is advised to visit the site and inspect any trial holes prior to submitting a tender and information is given in the Bills as to where the drawings and Conditions of Contract can be inspected if these have not already been provided.

The site visit

The estimator must visit the site of the proposed works to acquaint himself with local conditions prior to submitting a tender. The main points to be noted are:

(a) The position of the site in relation to road and rail transport facilities.
(b) The names of local and statutory authorities.
(c) Topographical details of the site. Whether the contours are such as would prevent the use of certain types of plant.
(d) Accessibility of the site. Mode of entry and any restrictions on unloading of materials.
(e) Ground conditions. The nature of the subsoil. The water table and its affect on excavation outputs. The amount of pumping required to keep the site free of water.
(f) Planking and strutting. The type of boarding required, open, close or sheet piling.
(g) The availability of water and electric mains. Their location will determine the length of temporary plumbing and wiring required to provide water and electricity to the site.
(h) Security. The requirements for temporary fencing and watchman.
(j) The labour situation in the area.
(k) The length and type of temporary roads required.
(l) The most suitable location for site offices and temporary buildings.
(m) Whether adequate space is available on site for the installation of the concrete mixing set-up.

Viewing the drawings

If copies of the drawings relating to the project have not been provided with the tender documents an early visit should be made to the Architect's office.

The main points to be noted are:
(a) The location of the site relative to adjacent or abutting buildings. The availability of working space. The availability of off-loading facilities if the site is in a congested area.
(b) Drawings and reports. Tracings should be taken of relevant details from the drawings. Reports of any soils investigations carried out should be examined.
(c) Scaffolding. The amount and type of scaffolding required internally and externally during construction.
(d) Hoardings, fans and gantries. The extent and type required.

Details of Preliminaries

Preliminaries to be considered for each tender are listed in the Standard Method of Measurement. The following notes amplify the headings therein:

1 *Plant, tools and vehicles*

The cost of any plant, tools and vehicles which cannot be readily allocated to specific bill items should be priced in the preliminaries as a lump sum. A mechanical hoist, for instance, handles many materials and the correct allocation of its time to each item in the Bills is impracticable.

Tower cranes, hoists, saw benches, power-driven pumps and general site transport are all included under this heading.

Small tools, wheelbarrows, picks, shovels, ladders, lamps, ropes, etc., are also included here, generally as a percentage of the builder's work.

2 *Safety, health and welfare*

The cost of:
(a) Site messroom for meals
(b) Sanitary accommodation
(c) Washing facilities
(d) Shelter from inclement weather
(e) Protective clothing, wet weather protection, boots, donkey jackets and protective headgear
(f) Drinking water
(g) First aid equipment

3 *Setting out the works*

The cost of an engineer and chainman if the works are of such a nature as to require specialist setting out. The use and waste of instruments.

PRELIMINARIES

4 *General foreman*

The cost of site staff to administer the project, including the following:
- Site manager or agent
- General foreman
- Cashier
- Timekeeper
- Checker
- Storeman
- Bonus surveyor
- Progress engineer
- Concrete engineer
- Quantity surveyor
- Safety officer
- Typist/telephonist

5 *Transport of workmen to site*

The cost of travelling time and fares paid to operatives. A proportion of the total number of men required may be available from local sources. Others will be paid travelling time and fares and the remainder may need to be imported from outside the district and paid subsistence allowance.

The statutory payments are laid down in the local *Working Rule Agreements*.

6 *Safeguarding the works*

The cost of a watchman for 7 evening shifts plus 2 shifts for Saturday and Sunday. Alternatively, a quotation can be obtained from a Security Company for periodic visits and checks.

7 *Water for the works*

The charge for providing water for building purposes must be obtained from the local water board. The cost may be quoted as a percentage of the contract value or as a rate to be paid on a metered supply.

The cost of providing temporary plumbing, standpipes, hoses and barrels is also included.

8 *Lighting and power for the works*

The cost of lighting the building internally or floodlighting the site during winter construction. Provision of power to small tools and the cost of installing temporary electricity cables for tower cranes, hoists, etc.

9 *Insurance*

The cost of insuring the works against fire and materials and plant against damage or theft.

10 *Temporary roads*

The cost of providing access for vehicles to and upon the site by the provision of temporary roads of hardcore, ashes or sleepers. Their subsequent removal and making good must also be included.

11 *Temporary buildings*

The cost of erecting, maintaining and finally dismantling and clearing away offices for site staff, stores, huts and other temporary buildings. Office accommodation for the Architect or Clerk of Works if required, is included under this heading.

12 *Telephones*

The cost of installing site telephones and the estimated weekly charge for calls made by the contractor. The cost of calls made by the Clerk of Works is normally covered by a Provisional Sum in the Bills of Quantities.

13 *Hoardings and gantries*

The cost of temporary hoardings or fencing to enclose the site and gantries for the unloading of materials, if required.

14 *Scaffolding*

The cost of all external and internal scaffolding required for the works.

15 *Drying out*

The cost of drying out the building prior to handing over. This involves heating the various rooms either by providing fuel for the permanent heating system or by using portable heaters.

16 *Cleaning the site*

The cost of regularly removing rubbish during the contract period and the cost of cleaning floors and windows and leaving the site tidy prior to the final handing over.

17 *Firm price addition*

Tenders can be invited on either a firm price or a fluctuating basis. If the latter, escalation of costs of labour and materials will be recoverable as a net amount. If the tender is required on a firm price basis, allowance must be made within the tender sum for future labour and materials increases until the completion of the contract.

A study of the published data regarding increases over past years is informative, but many building companies maintain their own cost records for use in compiling firm price additions.

Table of increased costs 1963–70

	'63	'64	'65	'66	'67	'68	'69	'70
Labour— gross rates	100	107	113	120	136	147	155	164
Materials	100	103	106	109	109	115	120	126
Building costs	100	105	109	115	119	126	133	140

It will be noticed that in 7 years labour costs have increased by an average of 9% per annum, materials by an average of 4% per annum and building costs by an average of 6% per annum.

It is also evident that costs are rising more rapidly in the later years.

Preliminaries

The preliminaries for a contract of about £250,000, to be completed in 52 weeks, are built up as follows:

The total of the Bills of Quantities is first of all analysed into its constituents:

	£
PC sums nominated subcontractors and suppliers =	70 000
Provisional sums =	8 000
Builder's subcontractors =	36 000
Builder's work =	100 000
Net total =	214 000

EXAMPLE

Contract: £250 000   *Time:* 52 weeks

1. *Plant, tools and vehicles*   £

   Hoists

	£		
1000 kg goods hoist per week	= 12		
Fuel oil and grease per week	= 3		
Operator—gross cost per week	= 35		
	= 45	× 20 weeks =	900

	£		
Small scaffold hoist per week	= 4		
Fuel per week	= 1		
Operator—part time per week	= 10		
	= 15	× 16 weeks =	240

   Pumps

	£		
75 mm pump per week	= 4		
Fuel per week	= 3		
Hose per week	= 3		
Operator—part time per week	= 5		
	= 15		
× 6 weeks × 4 pumps		=	360

   Small tools

1% on builder's work £100 000	=	1000

   Tower crane
   A typical build up for a tower crane is shown here but is not added to these preliminaries.

   Installation   £

(a)	Prepare ground and lay track base	=	250
(b)	Track materials	=	250
(c)	Transport to and from site	=	400
(d)	Lay track, erect and test crane	=	500
(e)	Dismantle crane and track	=	300
		=	1700

   Running costs

		£		
(a)	Hire per week	= 150		
(b)	Electricity, oil, grease	= 10		
(c)	Driver gross cost	= 46		
	Slinger gross cost	= 32		
		= 238	=	2380
			=	£4080

                                C/F 2500

Note that the installation of a tower crane forms a considerable proportion of the total cost.

				£
2	*Welfare*		B/F	2 500

Site messroom	£		
52 weeks @ £5 per week	= 260		
Erect and dismantle	= 100		
	= 360		
Latrines 4 @ £30 each	= 120		
Elephant shelter	= 25		
Protective clothing 26 men @ £5	= 130		
First aid equipment	= 20	=	655

3 *Setting out* £

Engineer 4 weeks @ £40	= 160		
Chainman 4 weeks @ £15 (part time)	= 60		
Instruments, pegs, use and waste	= 30	=	250

4 *Site staff* per week £

Site Manager	= 60
Cashier/Timekeeper	= 35
Storeman	= 30
Bonus Surveyor	= 25
Typist/Telephonist	= 10
	= 160

× 52 weeks =   8 320

C/F 11 725

			£
		B/F	11 725

5 *Travelling time and fares*

Value of builder's work £100 000
Labour = 40% = £40 000
Average weekly wage £30
 £40 000 ÷ £30 = 1333 man weeks
Average number of men = 1333 ÷ 52 weeks
 = 26

	£		

40% of operatives local—nil —
20% of operatives travel 5 miles daily
 and are paid £0·50 per week
 1333 man weeks × 20% × £0·50 = 133
30% of operatives travel 10 miles daily
 and are paid £1·50 per week
 1333 man weeks × 30% × £1·50 = 600
10% of operatives receive subsistence
 allowance at £0·85 per
 night £
 7 nights @ £0·85 = 5·95
 Periodic travel home
 10% = 0·60
 = 6·55
 1333 man weeks × 10% × £6·55 = 873 = 1 606

6 *Watching*

9 shifts per week @ £2·80 = £25·20 × 52 weeks = 1 310

7 *Water for the works* £

Temporary piping, barrels, hoses = 200
Water charge £0·25% of contract
 sum of £250 000 = 625 = 825

8 *Lighting and power* £

Temporary cable and fittings = 150
Electricity 52 weeks @ £10 per week = 520
Electrician = 150 = 820

		C/F	16 286

PRELIMINARIES 89

9 *Insurance*   £
                                                                                           B/F 16 286
  Third Party and Employer's Liability included
  in labour rates.
  Fire Insurance £0·15% per annum   £
  on contract sum of £250 000   = 375
  Plus 10% for professional fees   = 38   =   413

10 *Temporary roads*

   400 m² Hardcore and ash road @ £1 m²   =   400

11 *Temporary buildings*

   Office   £
   52 weeks @ £10 per week   = 520
   Erect and dismantle   = 150
   Stores and joiner's shop
   52 weeks @ £4 per week   = 208
   Erect and dismantle   = 60
   Cement shed
                                         = 40
   Site huts 2 @ £30 each   = 60   =   1 038

12 *Telephones*   £

   Installation   = 25
   52 weeks @ average £3 per week   = 156   =   181

13 *Hoardings*

   100 m long × 2 m high @ £1 per m   =   100

14 *Scaffolding*

NOTE

The cost of scaffolding varies according to the type (putlog, independent or birdcage), the height and the length of time it is required.

The range of cost of external scaffolding is from £0·40 per m² at 6 m high to £1·50 per m² at 30 m high.

Internal birdcage scaffolding costs £0·75 per m³ and is required to staircases and to liftwells.

For fixing ceiling finishes, mobile towers are often used.

   External putlog scaffolding   £
   2000 m² @ £1·00 m²   = 2 000
   Internal birdcage
   600 m³ @ £0·75 m³   =   450
   Mobile towers   =   150   =   2 600
                                                      C/F 21 018

			£
		B/F	21 018

15 *Drying out*

　　Fuel for heating system　2 weeks @ £35 per week　= 　70

16 *Cleaning site*

　　6 men　1 week @ £30　　　　　　　　　　　　= 　180

17 *Firm price addition*

　　Labour　£40 000
　　Allow for 8% increase 6 months after start of work
　　Value of builders labour remaining　£15 000

		£	
£15 000 × 8%		= 1 200	
Materials　£60 000　Allow 2% average		= 1 200	
Preliminaries　Allow		= 　500	= 2 900
	Total of Preliminaries	=	£24 168

NOTE

The Preliminaries for each contract must be individually considered. There is no single percentage, or range of percentages which can be applied to arrive at the correct amount of Preliminaries. Each tender presents its own variables and items of risk.

17 Completion of the Tender

Having priced the Bills of Quantities and the Preliminaries it remains only to complete the tender by the addition of the required percentages for Establishment Charges and Profit.

Establishment Charges

Establishment Charges, or Head Office overheads, are the permanently recurring costs of a contractor's head office administration, including the depreciation of buildings, the cost of staff and headquarters personnel, office furniture, rates, light and heat, stationery, postage and telephone charges and the cost of providing capital to run the business.

Establishment Charges are calculated annually as a percentage of the turnover. If the annual turnover is £10 000 000 and the central administration costs £750 000 per annum, the Establishment Charges are $7\frac{1}{2}\%$. This percentage is added to every tender submitted.

Profit

The margin of profit required by a contractor depends on many varying circumstances—the type of work (housing, industrial or commercial), the contract value, the prevailing market conditions, the likely amount of competition and the relative degree of risk in the project.

If a contractor is fully committed for the ensuing months he may not be prepared to undertake new work other than at a comparatively high profit level. A rival contractor may be short of work, or desirous of opening up in the particular locality and may be prepared to reduce his profit margin accordingly.

If too high a profit is added the contract will be lost; if too low, the contract may be won but is likely to show an inadequate return when completed, or even a loss.

The decision as to the amount of profit to be added is a management function and is not that of the Estimator, who assumes at this juncture a purely advisory role.

The amount of profit to be added is therefore arrived at by the manager by weighing the following relevant factors:
How full is the order book?
Is the contract attractive?
Is it too big, too small?
Is it the right sort of work?
Is labour available?
Is suitable staff available?

Is the Client/Architect/QS known from previous contracts?
How much capital is required?
Are there any onerous contract conditions?
What are the specific risks involved?
What is the effect on turnover?
It is the inclusion of profit which converts an Estimate into a Tender.

Having decided on the level of profit required, the overall tender figure can be calculated and is set out thus:

Tender summary

		£
Net total of Bill of Quantities		214 000
Preliminaries		24 168
		238 168
Establishment charges	$7\frac{1}{2}\%$	17 863
		256 031
Profit 5%		12 802
	Tender total	£<u>268 833</u>

The over-riding sum

The difference between the net total of the Bills and the tender amount is called the over-riding sum. In the example above, it is £54 833, made up of Preliminaries £24 168, Establishment Charges £17 863 and Profit £12 802. There are various methods of allocating this sum to the Bills of Quantities, the options being as follows:
1 All the Bill rates are increased by a percentage.
2 The builder's own rates are increased by a percentage.
3 The over-riding sum is added as a percentage of the measured work on the Final Summary.
4 The over-riding sum is shown as a lump sum in the Preliminaries Bill.
5 Any combination of the foregoing.

It is in the allocation of the over-riding sum to the Bills of Quantities that the expertise of the Estimator comes into play. The gambits are innumerable, depending on the contract conditions, the quantities, the items of risk and many other factors specific to the particular tender. These intriguing aspects are, however, beyond the scope of the present text.

18 Approximate Estimating

There are several methods of estimating the approximate costs of buildings in the early stages of a development, the main ones being as follows:
1 By a rate per square metre applied to the floor area.
2 By a rate per cubic metre applied to the volume.
3 By approximate quantities.

Rate per square metre

The total floor area of the building is measured between the internal faces of the external walls and a rate per square metre is applied to the area so obtained. The appropriate rate is obtained either from current pricing books or, better, from company records of the costs of similar buildings.

Rate per cubic metre

The cubic content of the building is measured by taking the external dimension of the walls and multiplying by the height from the top of the foundations to half way up the pitched roof, or in the case of flat roofs, to 600 mm above roof level. The resultant volume is multiplied by a rate per cubic metre.

Approximate quantities

Approximate quantities are taken off and priced. This method takes longer than the previous two but will generally be more accurate if adequate outline drawings and a specification are available. This method calls for a knowledge of current prices on the part of the Estimator or Quantity Surveyor.

Typical build-up of an Approximate Estimate for a factory

		£
Main factory building 3000 m² @ £65 m²		= 195 000
Offices 300 m² @ £80 m²		= 24 000
Garages and Outbuildings 200 m² @ £25 m²		= 5 000
		= 224 000
External works	£	
Roads and car parking	= 8 000	
Fencing	= 3 500	
Drainage	= 8 000	
Landscaping and lighting	= 4 500	= 24 000
		= 248 000
Contingencies 10%		= 24 800
		= 272 800

The figure quoted would be rounded off, in this case to £275 000

Approximate prices for buildings

	Per square metre £
Multi-storey office buildings	= 90
Single or two-storey ditto	= 80
Factories	= 65
Schools	= 85
Local authority housing—semi-detached	= 50
Ditto —multi-storey	= 70
Private housing	= 80

19 The Examination Approach

No student can hope to pass an examination in Estimating, or any other subject for that matter, by relying entirely on his native wit to see him through. A certain minimum of fact and knowledge is essential, be it eventually spread ever so thinly throughout the answer paper. However, there are certain things which can be done and certain things which must not be done in tackling an examination paper and a knowledge of these will increase the candidate's chances of success.

Each year a scrutiny of the examination scripts submitted reveals the fact that many candidates fail not necessarily owing to a lack of technical knowledge, but owing to a mis-application of the fundamental principles of how to set about an examination paper. Many adopt a haphazard approach which fails to do them justice, the three major faults being untidy setting out of the answers, bad allocation of the available time and finally, sheer carelessness. In an Estimating examination, clear setting out, balanced allocation of time and care in manipulating the calculations are vital in achieving success.

The candidate should include in his pre-examination preparations a study of the question papers of the previous years. More can often be learned in this way regarding the scope of the paper than from the syllabus itself. If the student's studies cover the scope of the questions over the previous five years, they can be reasonably expected to cover next near's paper.

The candidate should prepare his plan of campaign before entering the examination room. He should be ready to work to a pre-planned procedure in a set time.

One of the surest ways to fail is to have a last minute rush to arrive on time, or, even worse, to arrive late. The two or three hours allotted to the question paper are, to the candidate, the most important hours of that particular day. He should use every minute of them. He should arrive at the examination room in good time. He should see that his pens are ready and working and that his scale rules, pencils and other equipment are to hand. Such an observation may sound naive to the point of innocence but it bears repetition.

'Read the question paper' is the heart cry of every examiner offering advice to candidates, but it is of paramount importance in an Estimating examination. The question paper should be read quickly right through to establish which questions to select if given a choice and once that decision is taken the first question to be answered should be read again—every word and every figure. Underline the operative words. Make side notes of the key data. Many candidates confuse tonnes and cubic metres in the prices given for sand and aggregates. It is essential to operate in the correct UNITS requested, be they lineal metres, square metres, cubic metres or kilogrammes. Elementary mistakes of this nature are made every year with unfailing regularity.

Never jump to conclusions. Under the stress of examination conditions one's power of perception and observation deteriorate. Candidates are apt to anticipate the expected question and fail to notice that it may vary slightly therefrom. The correct reading of the question is a prerequisite to the correct answer.

Do not answer more questions than the number requested. 'Answer any three from five' means not more than three. If a candidate answers four questions the last one will be disregarded even though it would have gained more marks than one of the other three. If there is time to spare use it for revision and checking of the answers.

Deal with the compulsory questions, if any, first. An examiner generally attaches specific importance to these and to answer the minor questions first and leave inadequate time for the compulsory questions is to court failure. Otherwise, the order of answering should be the quickest and simplest first.

It is important in an Estimating examination to programme the available time. If there are 5 questions to be answered in 3 hours starting at 9 a.m., prepare a time schedule on the following lines:

	Minutes	*Complete by*
Read question paper	10	9.10
Question 1 (Compulsory)	50	10.00
3	20	10.20
5	25	10.45
2	30	11.15
4	35	11.50
Margin	10	12.00

The few minutes spent in organising a work programme will be handsomely repaid in keeping abreast of the clock. When running behind schedule, being aware of the fact is the first requirement to making up time.

Do not spend a disproportionate time on the less important details of a price analysis. In answering a question on brickwork it is foolish to spend three-quarters of the available time in building up the price of mortar to the detriment of the other components of the rate. If time is running out use an average cost of mortar to arrive at an overall answer. More marks will be gained by so doing than by becoming bogged down on the contributory details.

Keep all figuring neat and orderly. Set out the extension of each calculation clearly in the money column so that in the final addition the figures are easily totalled. Too many candidates endeavour to add up crooked columns of figures with tragic results.

THE EXAMINATION APPROACH

Do not be reluctant to show the rough calculations and working out. Use a separate sheet of paper or draw a margin to the left of the answer paper and work on that. The examiner is interested in following the candidate's line of thought and checking where his figures have come from.

In calculating rates for, say, drain trench excavation, do not base the calculations on 10, 50 or 100 lineal metres of trench. Base on the unit requested, i.e. 1 lineal metre. Similarly, assume 1 square metre of partition blocks and so on. The predilection of estimating students to base their working on anything from 10 to 100 square metres of the item is a perpetual mystery to examiners, the inevitable arithmetical miscalculation being practically guaranteed. Operate wherever possible in the single unit as shown in the examples of price analyses in the text.

To summarise, the sequence of events on the day of the examination should be as follows:

1 Arrive in good time, fully equipped.
2 Read the whole question paper.
3 Decide which questions to answer if given a choice.
4 Decide the order in which to answer them.
5 Prepare the time scale programme.
6 Read each question, every word and figure, before answering it.
7 Check that all the information given in the question has been used and that the answer is in the correct unit of measurement.
8 Check all arithmetic calculations.

20 Examination Papers

THE INSTITUTE OF QUANTITY SURVEYORS

THIRD/DIRECT MEMBERSHIP EXAMINATION 1971

Estimating and Analysis of Prices

WEDNESDAY 17 MARCH 1971 (9.30 am–12.30 pm)

TIME ALLOWED – 3 HOURS

**Answer ONE question from Section 1 and FOUR from Section 2
All questions carry equal marks**

NOTES

A In answering the paper the candidate is to assume that the questions relate to a ten-storey office block with a three-storey block forming the top of a tee plan shape, having an *in-situ* reinforced concrete frame with precast cladding, asphalt roof and open plan office areas. Tank rooms on the roof and a basement to the tall block. The water table is two feet below ground level.

The Conditions of Contract are the RIBA Agreement and Schedule of Conditions of Building Contracts, Private Edition (with Quantities) with Clause 31A deleted. Value of the contract is about £500 000, the contract period 18 months. The site is very confined with an area of about 60 feet by 40 feet available for temporary buildings and material storage.

Value of Main Contractors' work is £110 000 net, Builders Sub-Contractors £20 000 net, and Nominated Sub-Contractors, Suppliers and Provisional Sums £280 000 net (preliminaries, overheads and profit to be added).

B The gross hourly rates to be used in the price build-ups are as follows:

                Craftsmen   £0·74
                Labourers   £0·64

C Add 15% to all rates to cover Establishment Charges and Profit.

SECTION 1—One question to be attempted

1 Prepare the preliminaries on the proposed contract for any THREE of the following showing the detailed calculations:
(a) Site Staff.
(b) Temporary Plumbing and Electricity.
(c) Firm Price Addition.
(d) Plant (excluding scaffolding).

2 (a) Describe which items of labour cost are included in the gross labour rate.
   (b) Describe which items of labour are best priced in the Preliminaries and give reasons.

SECTION 2—Four questions to be attempted

3 Build up prices for the following:
   (a) Sawn formwork to horizontal soffits of suspended floors including strutting over 3·50 but not exceeding 5·00 metres high (assume 15 uses).
   (b) Wrot formwork to columns (assume six uses).
   (c) Sawn formwork to sides or vertical or battering walls (assume 12 uses).

4 Build up prices for the following:
   (a) Prepare and apply two coats of bituminous paint to concrete columns exceeding 300 mm wide.
   (b) 12 mm × 150 mm bituminous strip expansion joint including formwork in concrete floor.
   (c) Mastic pointing around window frames.
   (d) Bush hammer surface of concrete walls to expose aggregate.

5 Build up prices for the following:
   (a) Prepare, knot, prime, stop and paint two undercoats and one hard gloss finishing coat on general surfaces of woodwork (internally).
   (b) Touch up primer, paint two undercoats and one hard gloss finishing coat on general surfaces of metal (internally).
   (c) Prepare and apply one sealing coat and two coats emulsion applied by brush to concrete walls (internally).

6 Build up prices for the following:
   (a) 150 mm diameter cast iron drain pipe laid and jointed in trench, the joints made in gaskin and molten lead.
   (b) 12 mm diameter copper service pipe including all capillary joints in the running length fixed to fair faced concrete walls with and including plastic clips at 2·00 m centres.
   (c) 12 mm Bibtap including union to copper.

7 Give the outputs for the following indicating any labour assistance where applicable to tradesmen:
   (a) Excavate trench for boundary wall not exceeding 1·50 m deep.
   (b) Unload, handle and stack bricks off lorry to ground.
   (c) One brick wall in facings in cement mortar (1:3) in English Garden wall bond in boundary wall.

(d) 6 mm Georgian wired cast glass in panes exceeding 0·10 but not exceeding 0·50 square metres to wood with screwed beads including taking off and refixing beads.
(e) 102 mm half-round PVC eaves gutter on and including brackets to softwood.
(f) 150 mm × 150 mm × 6 mm white glazed tile as splash backs fixed with adhesive to concrete wall and pointed in white cement.
(g) 36 mm Granolithic (2:5) pavings in one coat to landings finished with and including carborundum dust trowelled smooth.
(h) Unload, handle and fix metal window size 1·00 m × 1·00 m to softwood sub-frame (measured separately).
(i) Fix only 150 mm flush bolt to softwood including mortice in concrete.
(j) Fix only flush door size 40 mm thick, size 560 mm × 1436 mm to softwood.

BASIC PRICES

Plant including Fuel (excluding Labour)
Pneumatic Hammer	£0·06 per hour
Props and Clamps	£0·10 per month each

Materials delivered site
Timber	£35·00 per cubic metre
Nails	£0·14 kg
Bituminous Paint	£0·16 per litre
12 mm × 150 mm Bituminous Strip	£0·10 per metre
Mastic	£0·16 per cartridge
Priming Paint	£2·25 per 5 litres
Knotting	£0·30 per 500 millilitres
Undercoat Paint	£2·40 per 5 litres
Gloss Paint	£2·35 per 5 litres
Emulsion Paint	£1·40 per 5 litres
150 mm Cast Iron Pipe	£2·75 per metre
12 mm Copper Pipe	£0·50 per metre
12 mm Bibtap	£1·64 each
Plastic Clips	£0·03 each
12 mm Capillary Fittings	£0·15 each
Lead	£0·75 kg
Gaskin	£0·14 kg

THE INSTITUTE OF QUANTITY SURVEYORS

Third/Direct Membership Examination 1971

Estimating and Analysis of Prices

WEDNESDAY 20 OCTOBER 1971 (9.30 am–12.30 pm)

Time Allowed—3 Hours

**Answer ONE question from Section 1 and any FOUR questions from Section 2
All questions carry equal marks**

Notes

A In answering the paper, the candidate is to assume that the questions relate to a large hotel extension. The development comprises a four storey bedroom block in traditional construction with strip foundations, brick loadbearing walls and partitions, concrete floors and a felt covered timber roof, together with a single storey service block of reinforced concrete frame, brick cladding and a felt covered timber roof. The ground is firm clay with the water level 1·00 m below existing ground level.

The Conditions of Contract are the RIBA Standard Form of Building Contract, Private Edition (with Quantities) with Clause 31A deleted.

The value of the contract is approximately £600 000 with a contract period of 15 months. The site is level with good access and space for site offices, temporary buildings and material storage areas.

The value of the Main Contractor's work is £150 000, Builders Sub-Contractors £40 000 net, and Nominated Sub-Contractors, Suppliers and Provisional Sums £300 000 (Preliminaries, Overheads and Profit excluded).

B The gross hourly rates to be used in the price build-ups are as follows:

         Craftsmen   £0·90
         Labourers   £0·75

C Price build-ups should be net with no addition for Establishment Charges or Profit.

SECTION 1

1 Describe the procedure an Estimator should adopt on visiting the site prior to tender and list the relevant points he should consider.

2 Indicate the method of building up prices for the following Preliminaries items in the contract:
   (a) Pumping and De-watering.
   (b) Temporary Electricity.
   (c) Hoists.
   (d) Contractor's offices, canteen, compound and latrines.

SECTION 2

3 Build up prices for the following:
   (a) Excavate trench for 150 mm drain pipe not exceeding 1·50 m deep commencing at ground level and average 1000 mm deep including grading bottom, planking and strutting, part return fill and ram and remove surplus average 50 m and spread and level over site. per m
   (b) Extra over for breaking out reinforced surface concrete bed 125 mm thick. per m

4 Build up prices for the following:
   (a) Portland cement concrete 1:3:6 (20 mm aggregate) in foundations over 150 mm but not exceeding 300 mm thick. per cu m
   (b) Ditto in 125 mm thick bed. per sq m
   (c) Heavy duty building paper laid on hard core (measured separately). per sq m

5 Build up prices for the following:
   (a) Precast concrete paving slabs 600 mm × 600 mm × 50 mm thick with 6 mm straight joints both ways bedded in cement and sand (1:3) laid to falls, cross falls and slopes not exceeding 15 degrees from horizontal. per sq m
   (b) Joint of paving slabs with other pavings. per m
   (c) Curved cutting. per m

6 Build up prices for the following:
   (a) 1 brick wall of common bricks laid to English bond in cement mortar (1:3). per sq m
   (b) Rough raking cutting. per m
   (c) Cut tooth and bond 1 brick wall to existing 275 mm cavity wall. per m

7 Build up prices for the following:
   (a) Two layer felt roofing finished with 10 mm mineral chippings bedded in bitumen in flat coverings to slopes not exceeding 10 degrees from horizontal.   per sq m
   (b) Ditto in coverings to kerb 200 mm girth with one fair edge.   per m
   (c) Make good around roof outlet.   No.

BASIC PRICES

Plant Hire Rate including fuel (excluding labour)

14/10 Mixer	£0·75 per hour
Dumper ($\frac{3}{4}$ YC)	£9·00 per week
J.C.B. 3	£1·55 per hour
Compressor 4 Tool	£0·60 per hour

Materials Delivered Site

Portland cement	£8·25 per tonne bagged
Concrete sand	£1·47 per tonne
20 mm Aggregate	£1·55 per tonne
Heavy duty building paper	£5·92 per 100 sq m
50 mm Precast paving slabs	£0·71 per sq m
Brickwork sand	£1·60 per tonne
Common bricks	£12·70 per 1000
Planking and strutting	£0·20 per sq m
Felt 12 m × 1 m	£2·00 per roll
10 mm Mineral chippings	£10·00 per tonne
Bitumen	£0·20 per litre

21 Basic Prices of Materials

	£
Aggregates and sand	*Per m³*
20 mm shingle	= 2·24
Washed sand	= 2·40
Hardcore	= 1·30
Ashes	= 1·50
Cement	*Per tonne*
Portland cement	= 7·90
Rapid hardening cement	= 8·42
Ready-mixed concrete	*Per m³*
1:2:4 20 mm aggregate	= 6·50
1:3:6 40 mm aggregate	= 6·10
Bricks	*Per thousand*
Commons	= 9·75
Sand-faced facings	= 13·75
Staffordshire blue	= 43·80
Engineering bricks, class A	= 31·50
Reinforcing rods	*Per tonne*
Mild steel rods 25 mm	= 56·25
12 mm	= 60·00
6 mm	= 65·50
	Per m²
Mesh fabric 6·17 kg per m²	= 0·40
Concrete blocks	*Per m²*
Hollow concrete blocks 50 mm	= 0·45
100 mm	= 0·70
150 mm	= 1·10
Damp proof course	*Per m²*
Bituminous felt dpc	= 0·25
Timber	*Per m³*
Carcassing timber	= 28·00
Joinery timber	= 38·00
Hardwoods: African mahogany	= 74·00
Oak	= 120·00
Teak	= 200·00

BASIC PRICES OF MATERIALS

	Per m^2
20 mm plain edged roof boarding	= 0·60
25 mm tongued and grooved floor boarding	= 0·80
12 mm insulation boarding	= 0·30

Metals — *Per tonne*
Rolled steel joists, channels = 60·00
Angles, tees = 65·00

Slates and tiles — *Per thousand*
510 × 255 mm slates = 132·00
Machine-made plain tiles = 18·75

Plumbing goods — *Per m*
75 mm plastic half round gutter = 0·25
75 mm cast iron rain water pipe = 0·70
20 mm lead piping = 0·48
25 mm copper piping = 0·74

Drain pipes — *Per m*
100 mm salt glazed stoneware pipes = 0·30
150 mm ditto = 0·45
225 mm ditto = 0·75

Each
100 mm bend = 0·28
150 mm ditto = 0·40
225 mm ditto = 1·00

Plaster — *Per tonne*
Hydrated lime = 8·25
Gypsum plaster = 12·90

Per m^2
10 mm plasterboard = 0·18

Paint — *Per 50 kg*
Washable distemper = 6·50

Per litre
Emulsion paint = 0·58
Primer = 0·58
Gloss paint—undercoat = 0·75
   ditto   —finishing coat = 0·80

Glass	*Per m²*
2 mm sheet glass	= 0·80
3 mm ditto	= 0·92
4 mm ditto	= 1·38
3 mm obscured glass	= 1·05
3 mm reeded glass	= 1·38
6 mm Georgian wired glass	= 1·50
6 mm polished plate	= 3·36

Pavings	*Per m²*
50 mm paving slabs	= 0·90

	Per m
125 mm × 250 mm precast concrete kerb	= 0·50

Rates for plant including fuel

	Per hour £
Mechanical excavator (0·53 m³)	= 3·71
5 Tonne tipper lorry	= 1·55
Tractor scraper (7 m³)	= 5·50
Crawler tractor shovel (1 m³)	= 2·75
10/7 Concrete mixer	= 0·33
14/10 ditto	= 0·40
5/3½ Mortar mixer	= 0·18

Index

Aggregates:
  all in, 28
  gravel, 28
  prices, 104
  quantities, 29
  sand, 28
  weight, 28
Analysis of prices:
  brickwork, 45
  carpentry, 58
  concrete work, 33
  drainage, 79
  excavation, 20
  formwork, 40
  glazing, 74
  joinery, 61
  painting, 76
  plasterwork, 69
  plumbing, 66
  precast concrete, 73
  preliminaries, 85
  reinforcement, 37
  roofing, 54
  rubble walling, 51
Approximate:
  costs of buildings, 94
  estimating, 93
Architect:
  drawings, 81
  site office, 84
Architrave, 60
Asbestos cement sheeting, 58
Ashes, 27

Banksman, 17
Barrel bolts, 63
Barrow wheeling, 14
Battens—roofing, 53
Bituminous damp proof course, 48
Blocks—concrete, 49
Boarding:
  floor, 61
  roof, 57
  wall, 58

Bolts, fixing, 63
Bricklayer:
  labour assisting, 42
  labour rate, 42
Bricks:
  common, 45
  English bond, 45
  facings, 45
  Flemish bond, 45
  metric, 42
  number, 42
  outputs, 42
  prices, 105
  stretcher bond, 45
  unloading, 42
Brickwork:
  analyses, 45
  arches, 42
  battering, 42
  cement mortar, 43
  damp proof course, 48
  facings, 45
  fair face, 46
  labour outputs, 42
  manholes, 42
  overhand, 42
  panels, 42
  raking out joints, 47
  scaffolding, 43
  underpinning, 42
Brushes, paint, 76
Building costs, 85
Bulking of spoil, 13
Butt hinges, 63

Carcassing timber, 57
Carpenter:
  labour assisting, 57
  labour rate, 57
Carpentry:
  analyses, 58
  labour outputs, 57
Casements, glazing, 74
Cast iron rain water goods, 66

INDEX

Cement, 28
Cement mortar, 43
Chalk, excavating, 13
Charges, establishment, 91
Cleaning site, 90
Code of Estimating Practice, 11
Columns, formwork, 39
Concrete:
  aggregates, 28
  beds, 31
  blocks, 49
  drainpipes, 79
  formwork, 38
  hand mixing, 30
  materials, 28
  mixers, 30
  mixes, 28
  placing, 31
  precast paving, 73
  ready mixed, 32, 105
  reinforcement, 36
  sand, 28
  transporting, 30
  vibrating, 31
Concretor:
  analyses, 33
  labour outputs, 31
Copper piping, 65
Cork tiling, 72
Cost:
  of buildings, 94
  of labour, 11
  of materials, 105
Cranes, tower, 86
Curing concrete, 32

Damp proof courses, 48
Disposal of excavation, 13, 18
Distemper, 76
Drain pipes, 79
Drain trenches, 78
Drain layer:
  analyses, 80
  labour outputs, 79
Drying out buildings, 90

Eaves gutter, 65
Emulsion paint, 76
Escalation:
  building costs, 85
  labour, 90
  materials, 90
Establishment charges, 91
Estimates, approximate, 93
Examination:
  approach, 95
  papers, 98–104
Excavating plant, 14
Excavation and Earthworks:
  analyses, 20
  disposal, 18
  hand outputs, 13
  machine outputs, 17

Facing bricks, 45
Fair face and point, 46
Fares, 83, 88
Felt roofing, 55
Fire insurance, 89
First aid, 87
Flooring, 72
  labour outputs, 72
Formwork:
  analyses, 40
  fabrication, 39
  fixing and striking, 39
  plywood, 39
  timber, 39
  wrought, 39

Glazing:
  analyses, 74
  labour outputs, 74
Guaranteed time, 9
Gypsum plaster, 68

Hand excavation, 13
Hardcore, 27
Hardwood, 60

INDEX

Head office overheads, 91
Hoardings, temporary, 84
Hoists, 32
Holidays with pay, 10
Housing, cost of, 94
Hutting, site, 89
Hydrated lime, 68

Inclement weather, 9
Increases in building costs, 85
Independent scaffolding, 89
Insurance:
   employers' liability, 11, 89
   fire, 89
   national, 10
Ironmongery, labour outputs, 63

Joiner, labour rate, 60
Joinery, 60
   analyses, 61
Joists, rolled steel, 64

Kerbs, 73
Knotting, 76

Labour:
   gross hourly rates, 11
Latches, 63
Lead:
   piping, 67
   sheet, 56
Lighting, temporary, 88
Lime:
   hydrated, 68
   mortar, 70
Linoleum, 72
Locks, 63
Lodging allowance, 88

Machine excavation, 14

Market prices of materials, 104
Masonry, 51
Materials:
   basic prices, 104
   waste, 7
Mechanical excavation, 14
   outputs, 17
   plant analyses, 17
   utilisation, 14
Mesh reinforcement, 37
Metal windows, 64
Metalwork, 64
Mild steel rods, 36
Mixers, 30
Mortar:
   analyses, 43
   cement, 43
   lime, 44

Nails:
   carpentry, 57
   roofing, 52
National Insurance, 8, 12
National Joint Council, 9
National Working Rule Agreement, 6, 9, 11

Oil paint, 76
Offices, site, 89
Overheads:
   head office, 91
   site, 81
Overtime, non-productive, 9

Painting and Decorating, 76
Partition blocks, 49
Paving slabs, 73
Payments—National Working Rule Agreement, 11
Pipes:
   concrete, 79
   copper, 67
   lead, 67
   salt glazed, 79

Placing concrete, 31
Planking and Strutting, 25
Plant costs, 16
Plasterboard, 69
Plasterwork, 68
Plastic:
    flooring, 72
    rainwater goods, 66
Plugging, 61
Plumbing, 65
    analyses, 66
    outputs, 65
Pointing, facings, 46
Portland cement, 28
Precast concrete, 73
Primer, 76
Preliminaries, 81
Prices of materials, 104
Profit, 91
Pumping, 86
Putty, 74

Quantities, approximate, 93
Quarry tiling, 72

Rainwater goods, 66
Rates of wages, 12
Ratio, hand to machine excavation, 19
Ready mixed concrete, 32
Reinforced concrete, 31
Reinforcement, 36
Rendering coat, 68
Rim locks, 63
Roads, temporary, 84
Roofing, 52
Roofing felt, 55
Rough rendering, 68
Rubber flooring, 72
Rubble walling, 51

Sand, 28
Sanitary fittings, fixing, 66

Scaffolding, 89
Schedule of basic prices, 105
Scrapers, 18
Screwing, 60
Selective Employment Tax, 10
Setting coat, plaster, 68
Setting out, 87
Shovels, mechanical, 14
Sheet:
    flooring, 72
    lead, 56
Sick pay, 9
Site preliminaries, 81
Skirtings, 62
Slabs, paving, 73
Slating:
    analyses, 54
    labour outputs, 53
Small tools, 86
Soldered joints, 67
Spoil, disposal of, 14, 18
Structural steelwork, 64
Storeman, 83
Sub-contractors:
    builders, 85
    nominated, 85
Subsistence allowance, 88
Surface finishes to concrete, 31

Telephone, site, 84
Temporary:
    offices, 84
    roads, 84
    water, 83
Tender, 92
Tiling:
floor, 72
    roof, 54
    wall, 71
Timber:
    carpentry, 57
    formwork, 39
    joinery, 60
    planking and Strutting, 25
Tool money, 6

INDEX

Travelling allowances, 83, 88
Trench excavation, 78
Turnover, 91

Undercoats, 76
Unloading, 7
Utilisation of plant, 14

Vibrating concrete, 31

Wages, build up, 12

Wall ties, 45
Wall tiling, 71
Waste, 7
Watching, 88
Water:
  cost, 88
  table, 81
  temporary, 83
Windows, 62
Wiped soldered joints, 67
Works on site, 82
Wrought formwork, 39